What Hashish Did To Walter Benjamin

Books by Sebastián Marincolo

High. Insights on Marijuana.

Dogear Publishing 2010

High. Das positive Potential von Marijuana.

Tropen/Klett-Cotta 2013

Blog on the Marijuana High

www.marijuana-insights.com

Homepage

www.sebastianmarincolo.de

What Hashish Did To Walter Benjamin

Mind-Altering Essays on Marijuana

Sebastián Marincolo

For
Lester Grinspoon

Mentor and friend,
who so generously shared his unbelievable knowledge with me
in so many conversations

What Hashish Did To Walter Benjamin.
Mind-Altering Essays on Marijuana

Library of Congress Cataloging-In-Publication Data

Marincolo, Sebastián 1969-
What Hashish Did To Walter Benjamin. Mind-Altering Essays on Marijuana
p. cm. – (What Hashish Did To Walter Benjamin)
Includes biographical references.
ISBN 978-3-9817712-0-6
1. Marijuana 2. Cannabis 3. Walter Benjamin

First edition: November 2015
Stuttgart, Germany
Publisher: Khargala Press
Number of Pages: 190
First Printing: 2015

Summary: 20 neurophilosophical essays about the positive potential of the marijuana high and how it positively affected luminaries such as Walter Benjamin, Carl Sagan, and Louis Armstrong.

Publisher's Note

We take great care to ensure that the information in this book is accurate and presented in good faith, but no warranty is provided or results guaranteed. This material is intended for informational purposes only. The publisher does not condone illegal activity of any kind.

Printed in the U.K.

Contents

What Hashish Did To Walter Benjamin

Foreword by Joe Dolce

In older, more spiritually oriented cultures, people who used plant substances to get closer to God relied on shamans to navigate the treacherous pathways of their journeys. Our forbearers might have been wiser than we are today. We use substances with little guidance, cannabis included. While cannabis isn't as mind bending as other chemicals, big changes have occurred in the plant and in the science surrounding it in the last 30 years. As stronger strains, plus new forms of concentrates and edibles come online, and as we learn more about the complex chemical factory within this plant, a little shamanic guidance might be in order. Sebastián Marincolo is the modern-day shaman I turn to first.

To be clear: the finest shamans aren't simply acting out of superstition. They're drawing upon a centuries-old body of systematic knowledge. Their skills are refined, their techniques honed and tested over time. Likewise, Marincolo's writing is based on a vast body of knowledge, but also represents a new methodology for researching the potential of the marijuana high. It is informed by the philosophy of mind, the cognitive sciences, psychology, chemistry, neurobiology and a systematic analysis of hundreds of anecdotal reports. It blends hard science with the warmth of human experience and a deep appreciation for the complexity of human consciousness.

While medical cannabis is capturing most of the attention at the moment, the vast majority of users – an estimated 85% - are interested in the high; first and foremost. Marincolo is pushing that even further. He wants to know how the high can spur creative thinking, deepen empathic understanding, enhance the ways we pay attention, or bring hidden memories to the fore. Cannabis has always been able to do some of these things, and in the third part of this book Marincolo illuminates how a wide range of artists and thinkers – John Lennon, Louis Armstrong, Carl Sagan, and yes, Walter Benjamin among them – have all used this contemptible weed to spur some of their finest ideas. Rather than dwelling on the supposed dangers of cannabis – none of which have ever been proven – he's asking a far more intriguing question: How can we use it to enhance our

At the same time a handful of thinkers and evolutionary biologists have started to ask why exactly such a plant has been such a reliable companion to human beings for over 10,000 years. One of the more intriguing theories comes from Dr. Allyn Howlett, the chemist who discovered the endocannabinoid system of receptors in the body with which the cannabis plant interacts. She postulated that after Adam and Eve were evicted from the garden of Eden the magic flowers helped them not only to survive, but to thrive in this newly hostile environment. After a grueling day on the savannah hunting for supper Adam might have nibbled some cannabis flowers at night to soothe his aching muscles and help him forget the trauma of being chased by that herd of wild boars. Likewise it could have helped Eve endure the pain of childbirth, and then to forget about it so she could continue producing more little ones.

In short: this plant likely helped our ancestors contend with the human condition. As scientists now suspect, cannabis decouples certain traumatic memories from the fear/stress response. Without being able to forget pain and the associated stresses, our forebearers would have reproduced and adapted much more slowly.

Although the stresses of the modern world have changed cannabis might be equally useful today. In the 24/7 information age, we need to find new ways of relaxing, thinking differently and staying connected to each other (rather than our screens). Cannabis isn't the only solution, but it's one that grows in the earth and is safer than any of the 500,000 prescription drugs currently on offer, most of which have a list of side effects longer than this book.

While science has taken us far in understanding the chemistry of cannabis it can only take us so far in understanding the magic of a high. In the same way that neuroscience can show that Bach and Taylor Swift both trigger activity in certain areas of the brain, science is utterly incapable of explaining why Bach plumbs the depth of our emotion while Taylor doesn't. To see beyond brain scans and synaptic firings, we need thinkers such as

Marincolo, to ask these Big Questions. Thankfully, he's also exploring a few of the more modern mysteries, such as, "Why is a vaporizer high different from that of a joint?" It's something that puzzles everyone who imbibes, but to which I could find no satisfying answer until I read the final chapter of this book. I suggest you do so immediately.

Marincolo's writing illuminates new paths in the ever-unfolding story of this plant. Will scientists be able confirm his theories? Possibly, but science is a slow and painstaking affair. While research grinds on I suggest you follow some of his shamanic guidance to increase the amount of fun you have with the magic weed and to explore new ways of investigating its potential. You'll never think about your high the same way again.

Part I Turn On

Cannabis, Mind Enhancements, and Culture

"(...) for the actual experience of the smoked herb has been completely clouded by a fog of dirty language by the diminishing crowd of fakers who have not had the experience and yet insist on being centers of propaganda about experience"

Allen Ginsberg, American Poet, (1926-1997)

"Unquestionably, this drug is very useful to the artist, activating trains of association that would otherwise be inaccessible, and I owe many of the scenes in 'Naked Lunch' directly to the use of cannabis."

William Burroughs, American writer (1914-1997)

A Distorted and Misinformed Perspective on Marijuana

After more than 80 years of an almost worldwide prohibition our outlook on marijuana and its mind-altering effects is mostly dominated by fear, ignorance, and disinformation. There are still dozens of myths circulating about the negative effects of marijuana; myths that have been purposefully created and spread for decades. Many activists have tried to argue against these myths and are fighting to legalize the use of marijuana; but even liberal minded activist are often wary of mentioning the positive potential of marijuana when it comes to its mind-enhancing effects. It is also a strategic decision to not talk openly about the positive mind-altering potential of cannabis. Political arguments are usually based along the lines of proving the incredible usefulness of cannabis as a medicine, or on arguing that prohibition is detrimental to our whole society. The strategy makes sense; both arguments are correct and important. Nevertheless, I think we should not remain silent when it comes to the many mind-enhancing uses of marijuana.

For thousands of years cultures around the world have used cannabis not only for nutrition, clothing, for medical and many other purposes, but also valued its mind-altering potential. In China, the country that was often described as "the land of mulberry and hemp" in classical times, recorded uses of hemp go back more than 10.000 years ago. According to legend, Emperor Shennong, the founder of Chinese herbal medicine, wrote the first pharmacopedia *Pên-ts'ao Ching* around 2700 B.C.. It recommends the uses of cannabis as a medicine for many purposes – for absent-mindedness, for example – and also mentions other psychoactive properties, stating that taken in excess, it will make a user 'see devils' and that 'if taken in excess will produce hallucinations'.[1] The Chinese culture did not appear to be too involved in the use of cannabis as a psychoactive plant, even though cannabis seemed to have important influences through shamanic practices and in Taoism:

The Chinese Emperor Shennong tasting plants.

"By the first century AD (...) Taoists were suggesting that cannabis seeds be tossed into one's incense burner in the direct knowledge that they would cause hallucinations. This time, as advocated by the shamans, the visions were desirable: such hallucinations, it was noted, offered a shortcut to immortality and at the same time would help the intoxicated person to see spirits. But in the event, cannabis smoking was never a major Chinese preoccupation."[2]

On the other side, the cultural history of India is heavily linked with the use of cannabis for mind-altering purposes. Cannabis was brought into India by the Aryans, who invaded from the North around 2000 B.C. The first mention of the use of *bhang* – a drink made with cannabis and other ingredients including pepper, ginger, cloves, cinnamon, almonds and nutmeg – for mind-altering purposes can be found in the *Vedas (veda is Sanskrit for "knowledge"),* a collection of originally orally transmitted songs. The *Athar-*

---

1 Compare Jonathan Green (2002), *Cannabis*, Thunder Mouth Press, New York, p.40.

2 *Ibid, p.41.*

vaveda states that bhang is on of the "*five kingdoms of herbs ...which release us from anxiety.*"[3]

India's religious tradition is deeply rooted in the use of cannabis as a psychoactive plant; the Hindu god *Shiva* ("The Auspicious One"), one of the most important gods, is also known as the 'Lord of *Bhang*'. *Sadhus* (holy men) in India have used cannabis for centuries to come closer to their deities during their meditations. *Sanskrit*, the primary liturgical language of Hinduism, has many names for cannabis all of which praise it: *vijaya* and *jaya*, ("victorious"), *virapattra* ("leaf for heroes"), *capala* ("the light hearted"), *ananda* ("the joyful"), *vakpradatava* ("speech giving"), *medhakaritva* ("inspiring of mental power").[4] The 17th century text *Rajvallabha* describes *bhang* as follows:

"*It creates vital energy, increases mental powers and internal heat, corrects irregularities of the phlegmatic humor, and is an elixir vitae. (...) Inasmuch as it is believed to give victory in the three worlds and to bring delight to the king of gods [Shiva], it was called [vijaya] victorious. This desire fulfilling drug was believed to have been obtained by men on earth for the welfare of all people. To those who use it regularly, it begets joy and diminishes anxiety.*"[5]

The use of *bhang* for psychoactive purposes was and still is widespread in India and is part of many cultural rituals and festivities. Cannabis also plays an important role in the Tantric religion, which evolved in India from around 500 A.D.

The Aryans also brought cannabis to Persia, where the prophet Zoroaster, who is said to be the author of the *Zend-Avesta,* the counterpart to the *Vedas* containing approximately two million verses, arguably also used it around 600 B.C. The book tells about two mortals drinking *bhanga* (proba-

---

3 Compare Ernest L. Abel (1980), *Marihuana. The First Twelve Thousand Years.* Plenum Press, New York.

4 Marincolo, Sebastián (2010), *High. Insights on Marijuana*, Dog Ear Publishing, Indianapolis, and Marincolo, Sebastián (2013) *High. Das positive Potential von Marijuana.* Tropen Verlag, Stuttgart.

5 Grierson, G. A. (1993-4), "On References to the Hemp Plant Occurring in Sanskrit and Hindi Literature," in *Indian Hemp Drugs Commission Report*, Simla, India 3: pp.247-8.

bly the Persian version of the Indian bhang) to go to heaven and have the highest mysteries revealed to them.[6]

Around the same time, other groups of Aryans who would later become known as the Scythians moved from central Siberia further west and up to northern Greece, claiming huge territories. Famously, according to the Greek historian Herodotus, their funeral rites included some kind of steam bath in tents, in which marijuana seeds were thrown on hot stones and the Scythians would howl in joy – clearly indicating that they ritually used cannabis for its psychoactive properties. Findings of the Russian archeologist S. I. Rudenko also suggest that the Scythians used cannabis in their everyday life. Although the Scythians disintegrated as an entity, their descendants, habits and their use of marijuana spread to Northern Europe, where we can still find traditional ritualistic use of marijuana in many countries such as Lithuania, Poland, Latvia, and the Ukraine.

French Writer Victor Hugo, 1802-1885

These are only the beginnings of the worldwide spread of the cannabis plant and its use not only for fibers, food or medicine, but also as a psychoactive plant used in many diverse rituals and practices. Cannabis went on to the Arabic world, where hashish used as a psychoactive substance has a strong influence on religion and society even today. As early as around 300 A.D., Arab traders brought cannabis to Africa – where it also had a huge influence on culture. From Egypt, Napoleon's defeated retreating troops brought their habit of hashish use back to France in 1801, and more than 40 years later, French intellectuals and artists like Théophile Gautier, Charles Baudelaire, Gerard de Nerval, Victor Hugo, and Eugène Delacroix dressed up in Arabian clothes in a beautifully decorated Pimodan House, which they rented for their "*Club des Hashashins*" to experiment with

---

6 Compare Abel, Ernest L. (1980), *Marihuana. The First Twelve Thousand Years.* Plenum Press, New York, p.22.

large doses of hashish marmalade for inspirational reasons. Some of their writings on their experiences with hashish as a psychoactive drug will turn out to be vastly influential later. Inspired by Baudelaire's writings on hashish, the widely influential philosopher and essayist Walter Benjamin will for instance start his own experimentation with hashish under the supervision of drug experts in 1927. In England, the society of the "Hermetic Order of the Golden Dawn" is formed in 1888, with its members also experimenting with hashish. Two of their most prominent recruits would be the infamous occulti, Aleister Crowley and the poet, W. B. Yeats.

The inspirational use of hashish or marijuana by millions of people around the world has left its mark on many cultures and cultural developments. In our recent history, it arguably played an eminent role in the development of musical traditions like jazz, reggae, rock, pop and electronic music, as well as in the revolutionary anti-war and global peace movements in the late 1960s.

This is only a small excerpt of some inspirational, religious and other uses of cannabis as a psychoactive substance in world history. Many important questions remain unanswered. How, exactly, did early shamans in various regions use cannabis for psychoactive and healing purposes? How did this influence early cultural developments? What was the role of the psychoactive use of cannabis in the development of religions like Hinduism, Tantric belief system, and Buddhism? How, exactly, did the experience of the altered state of consciousness during a high positively affect writers, philosophers, artists, filmmakers, scientists, musicians and others?

History as a discipline will need the help of modern science to address these questions more seriously. Certainly, the study of literary sources, new findings from archaeological sites and ethnological research will help to answer these questions, but first and foremost, we need to investigate scientifically the altered state of the cannabis high itself. How can a marijuana high influence a musician's performance? Can it actually help a writer or an artist in the creative process? Or lead to deep and meaningful insights and make users see patterns they have not seen before? In short, how does a cannabis high affect our mind?

Cannabis as a Cognitive Enhancer

It is time to openly address the issue and to talk about the mind- and life-enhancing potential of cannabis. What is the positive potential of the cannabis high as an altered state of consciousness? How can a high be used to enhance various thought processes and activities? How much can it mean to individuals? And how much of a positive impact did the use – not abuse – of cannabis as a mind-enhancer have on human culture and history?

I have argued in my two previous books on the cannabis high that it can lead to various systematic changes in cognition and perception, which can be positively used by skilled and knowledgeable users. Many cognitive enhancements have been described by a myriad of users in history over and over again: a hyperfocus of attention and a concentrated feeling of being in the here-and-now of experience, an intensification of sensual experiences, often experienced with a sharp analytical ability to distinguish even tiny nuances in perception, an enhancement of episodic memory and of our imaginative abilities; associative mind racing, enhanced creative thinking and better pattern recognition abilities – for all kinds of patterns, be it in nature, art, in science, or in the behavior of others. Furthermore, numerous users have reported an enhanced ability to introspect bodily states, and also to come to other introspective judgments regarding their mood and character. And we have countless detailed descriptions from marijuana users about the enhancement of empathic understanding during a high, as well as a generally enhanced ability to come to deep insights in all kinds of intellectual areas – since the time of Zoroaster, many 'mortals' have reported that they got high, "went to heaven, and had the highest mysteries revealed to them" – using a slightly different wording, though.

Set, Setting and the 'Robustness' of Cognitive Effects

As Harvard psychologist Timothy Leary and others have pointed out, the high (as well as psychedelic states induced by substances such as LSD or psilopsybin) heavily depends on the set and setting of the user, whereby the "set" for Leary included the preparation of the user, his personality structure and mood. The "setting" is the physical environment of use, but also the social environment of others present during the experience and the cultural environment.[7] That may sound a bit technical, but is easy to understand: the son of a catholic priest smoking his first joint under peer pressure in his college dorm in Muskogee, Oklahoma in 1954 will not experience such a relaxed and happy high as a liberally educated neo-hippie smoking a joint of the same size and quality at the Burning Man festival with his stoner friends in 2015, even if they smoke the same amount of marijuana.

The Temple of Joy at the Burning Man Festival 2002

Some of the more basic cognitive changes during a high are more robust and not as heavily dependent on set and setting, for instance, the hyperfocusing of attention or the intensification of various sensory perceptions. When you watch television while getting high, you will strongly focus on what you see and forget about reality; when you eat, you will focus intensely on your food, and when you have sex, you will have a stronger focus on your sensations and your partner. In other instances, you may just focus on an inner stream of thoughts, memories or images. In all those situations, the content of what you focus on may be different but your focus, your selective attention will always becomes stronger during a high.

Other cognitive enhancements during a high are more complex and more dependent on set and setting. Take for example the ability for introspection. Many users have described in various ways how a high helped them to come to introspective insights about their character. I am convinced that

---

7 Timothy Leary, Ralph Metzner, Richard Alpert (1964/1992), *The Psychedelic Experience. A Manual Based on the Tibetian Book of the Dead.* Citadel Press, Kensington.

cannabis can indeed affect cognition in a way that can help you to come to introspective insights; but you will probably not get valuable introspective insights when you decide to play an ego-shooter computer game every time you get high. If you want to use marijuana to come to introspective insights, you should be in a less distracted environment and be willing to reflect upon yourself in some way. I have therefore always emphasized that cognitive enhancements during a high do not come automatically; marijuana has the *potential* to lead to all those enhancements, but it needs skills, knowledge and the right attitude to experience the more complex enhancements of empathic understanding, introspection or insights. Users have to carefully choose their environment and they have to learn what dosage of which strain is good for them in a certain mood and for a certain kind of activity. Also, as I will argue later in this book, the means of administration plays a big role; a vaporizer holds a completely different potential for users than burnt joints.

The Many Positive Uses of the Marijuana High

Many cannabis users in history have managed to develop their skills to use cannabis under certain circumstances for various mind- and life-enhancing uses. In current discussions about the various uses of cannabis, these uses are often lumped together under the category of "recreational", or as "inspirational". In my view, these expressions definitely do not fully capture the immense spectrum of uses reported by cannabis users. Naturally, many users value cannabis for relaxation, for fun or other "recreational" activities; and, let us not understate the case, these uses are definitely extremely valuable to many. Similarly, artists, writers, musicians and others have used cannabis for "inspirational" purposes to produce art, to enjoy the view of a landscape or to generate all kinds of ideas. Again, this kind of use is incredibly important and has helped many to develop their art, to grow as a person or to generate valuable ideas in many areas of their lives. But if we take a closer look at how skilled users have managed to actually integrate and positively use a cannabis high in their lives in so many ways, it becomes clear that "recreation" and "inspiration" account for only a small percentage of the many positive uses and the vast potential of this plant.

During my research in the last 10 years I have read, heard and analyzed hundreds of accounts of users describing a huge range of uses of the enhancing effects of marijuana. Many users report how hashish or marijuana can get them in the "here-and now"; they explain how they feel more connected to others and to their own feelings, how their high led to a better understanding of their kids, of their partners, to deep conversations and family healing. They perceive more nuances in a painting, or suddenly expand their musical or artistic horizon. Numerous users value the ability to travel back in time during a high and to re-live experiences and to feel like a child again and they report how it helped them with ideas for problem solving or got them off an alcohol or other addiction. Others report in detail how it helped them to enjoy sex more and to be more empathic lovers. They become creative in cooking and come up with new desserts, they use a high to find recognize patterns in music, or introspectively find patterns in the way they lead their marriage or in the way they walk – and after this perception decide to eventually change their bad habits forever. They use the slowdown of time in their perception to dwell in an infinite moment of bliss swimming in thermal water. An illiterate describes how a high for the first time helped him to read full sentences, others describe how they understand a foreign language better. Let us take a look at some prominent marijuana users to shed some more light about how important the enhancing use of marijuana has been to them – and through them and their work, to millions of people.

Writers and Their Use of Marijuana

The American writer Norman Mailer won the renowned Pulitzer Price twice as well as the National Book Award. In an interview with the High Times, he said about marijuana:

"*I always tell my kids - I don't know if they listen or not - that what I think is, get their education first and then start smoking pot. At least there is something to run downhill with. Because what I find is that pot puts things together. Pot is marvelous for getting new connections in the brain. It's divine for that. You think associatively on pot, so you can have real extraordinary thoughts. But the more education you have, the more you have to put together at that point, the more wonderful connections there are to see in the universe."*[8]

Many other writers found marijuana helpful for their work for various reasons. The French poet Charles Baudelaire was part of the *Club des Hashischins* in Paris, where in the mid-19th century famous writers including Alexandre Dumas and Victor Hugo and other French intellectuals and artists would meet to experiment with large amounts of hashish marmalade. Whereas Mailer mentioned how good pot worked for him to make new associations, Baudelaire also described how fast these associations often come:

"But a new stream of ideas carries you away: it will hurl you along in its living vortex for a further minute: and this minute, too, will be an eternity, for the normal relation between time and the individual has been completely upset by the multitude and intensity of sensations and ideas. You seem to be living several men's lives in the space of an hour."[9]

Beat poet Allen Ginsberg, best known for his famous poem "Howl", used marijuana a lot and wrote extensively on its effects:

"(...) marijuana consciousness is one that, ever so gently, shifts the center of attention from habitual shallow purely verbal guidelines and repetitive second-hand ideological interpretations of experience to more direct, slower, absorbing, occasionally microscopically minute, engagement with sensory phenomena during the high moments or hours after one has smoked."[10]

The influential philosopher, literary critic and essayist Walter Benjamin was strongly influenced and inspired by Baudelaire and Marcel Proust who had also used cannabis. Benjamin experimented with Hashish and wrote several essays about his experiences. In my essay "What Hashish did To Wal-

---

8 In: Hager, Steven (ed.) (1994) High Times Greatest Hits. Twenty Years of Smoke in Your Face. St. Martin's Press, New York, p.66.

9 Baudelaire, Charles "The Seraphic Theatre", translated by Norman Cannon, in: David Solomon (ed.) (1966), *The Marijuana Papers,* Signet Books, New York, p.190.

10 Ginsberg, Allen "First Manifesto To End The Bringdown", In: *Deliberate Prose. Selected Essays 1952 –1995,* Edited by Bill Morgan (2000), New York: Harper Colins Publishers, p.87.

ter Benjamin", I will argue that contrary to the belief of many Benjamin interpreters, his essays on the effects of hashish contain brilliant observations and, even more importantly, that some crucial ideas in Benjamin's important work have been influenced by his use of hashish.

Many other writers used marijuana; according to *veryimportantpotheads.com*, the list includes Arthur Rimbaud, William Butler Yeats, Robert Louis Stevenson, Rudyard Kipling, Jack Kerouac ("On the Road"), Jack London and filmmaker Hal Ashby ("Harold and Maude"), to name only a few.

Artists and Musicians on Cannabis

Throughout history, countless artists and musicians have experimented with cannabis, many of them were long-time users and some explicitly reported how the cannabis high had helped them to work on their music or art.

"The Woman of Algiers", painting by Eugène Delacroix, 1834

The French painter Eugène Delacroix had made experiences with hashish, was also a member of the *Club des Hashishins* and known for his vivid imagination and his use of expressive colors, paving the way for expressionism. His famous painting "The Woman of Algiers" depicts Algerian concubines smoking a water pipe used for opium and hashish, a painting much admired by another painter who used hashish, Pablo Picasso.[11]

The influential Mexican painter Diego Rivera also used marijuana:

"The Book of Grass contains an account by the actor Errol Flynn telling how Rivera asked him whether he had ever heard music come from a painting. Then the artist proffered Flynn a marijuana cigarette, explaining, „After smoking this you will see a painting and you will hear it as well." Flynn tried it and had a fascinating experience, in which he heard the paintings 'singing'."[12]

As I will argue in my essay "Marijuana and the Early Evolution of Jazz", the marijuana high was crucially important for the evolution of jazz with its altered rhythmic structure based on an altered sense of time during a high and its creative, free flow improvisation on stage. The experience of the high was also central for other musical traditions such as reggae. Bob Marley, who smoked massive amounts of marijuana, knew about the potential of the plant also when it comes to introspection and insights. He once said: "*When you smoke the herb, it reveals you to yourself.*"

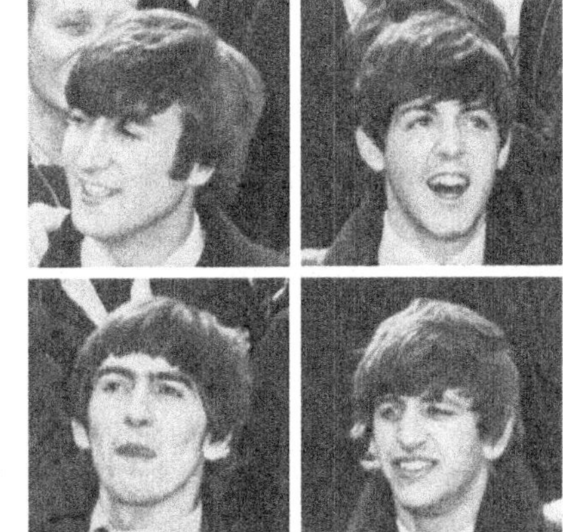

The Beatles, turned onto marijuana in 1964 by Bob Dylan, were strongly influenced by their use of marijuana; it helped them to open their minds and to get deeper involved in the thinking of the evolving counterculture of the 1960's:

"The crucial catalyst for the Beatles' transformation from lovable mop-tops to high-minded rebels was their involvement with consciousness-raising drugs, specifically marijuana and LSD. No one liked fun more than the Beatles, but for them drugs were not simply about having a good time. Marijuana and LSD

[11] Komp, Ellen (2014), www.veryimportantpotheads, entry on Eugène Delacroix, http://www.veryimportantpotheads.com/delacroix.html

[12] *Ibid.*, entry on Diego Rivera, http://www.veryimportantpotheads.com/rivera.html

were also and more profoundly tools of knowledge, a means of gaining access to higher truths about themselves and the world. Indeed, it was above all the "desire to find out," as Harrison later put it, that lay beneath their involvement not only with mind-expanding drugs but with Eastern philosophy as well. (...) It was marijuana that came first and triggered "the U-turn," as McCartney put it, in the Beatles' attitude toward life."[13]

Clearly, the marijuana high affected The Beatles, their music, and with it, hundreds of millions of people around the world – not just their everyday fans but bands, songwriters, artists, even politicians.

Of course, the list of prominent users does not only include writers, artists and musicians but also scientists, business people, comedians, actors, and many others who used marijuana for various enhancements that influenced their lives and work. Projects like Lester Grinspoon's *marijuana-uses.com* (where he collects reports and essays about positive enhancing uses of marijuana) and Ellen Komp's *veryimportantpotheads.com* are an important start for a better understanding of how much a whole spectrum of enhancing uses of marijuana has influenced our culture and society as a whole – how much we probably all owe to people who have used marijuana for various enhancements. However, if we want to arrive at a deeper understanding of the positive impact of these enhancements on our society, we have to investigate deeper. We have to free our minds from past disinformation campaigns and reconsider and research our cultural involvement with this outstanding plant.

In my research I have mainly tried to come to a better understanding of the effects of cannabis on higher cognitive functions, episodic memory retrieval, attention, pattern recognition, imagination, introspection, empathic understanding and insights. As our knowledge grows about the marijuana high we are beginning to understand why so many people have used cannabis for so many purposes, and how artists, writers, scientist, musicians and others benefited from this altered state of consciousness. I suggest that we

---

[13] Hertsgaard, Mark (1995) *A Day in the Life: The Music and Artistry of the Beatles.* Chapter 6: *We All Want to Change the World: Drugs, Politics, and Spirituality.* In: *Lester* Grinspoon *(ed.) marijuana-uses.com,* http://marijuana-uses.com/we-all-want-to-change-the-world-drugs-politics-and-spirituality-by-mark-hertsgaard/

take a closer look at the involvement of culturally influential personalities with marijuana (and, for a bigger picture, with other mind-altering substances) and reinvestigate their biographies to see how much their experiences and use of mind-altering substances like cannabis influenced their work – and with it, on all of us. The overview of personalities here is only the very beginning, and my essays on Walter Benjamin can also only be the beginning of such an investigation.

I hope to see more scientific projects in this direction soon. I firmly believe that future generations will come to a completely different understanding of what marijuana – as well as other psychoactive substances – have done for our culture, and I am convinced that its mind-altering potential should be considered at least as important to us as many of the other wonderful uses of cannabis as a medicine, a nutrient, as a resource for clothing, fibers and other useful products. I am fully convinced that the mind-altering potential of marijuana, once unleashed, could significantly help to heal a sick world – a point that I will argue in more detail in my essay on Carl Sagan.

Many of the positive influences of cannabis on society through the work and inspiration of artists, writers, scientists and others came during grim years of worldwide prohibition. Now, I have argued above that many of the marijuana mind-enhancements crucially depend on the skills of informed users and a favorable environment. So, let me conclude here with a question: Given how much marijuana has already positively influenced literature, art, music and many other cultural developments, how much could the skilled and careful use of cannabis do for our culture if we would legalize it and educate the public better as to its benefits?

Just Another Altered State of Consciousness

"Why does the eye see a thing more clearly in dreams than the imagination when awake?"
Leonardo da Vinci (1452-1519)

Altered States of Consciousness and Western Culture

There is the marijuana high and there is normal consciousness. And being high is like being a little confused, euphoric, and retarded.

Briefly, this is still the mainstream attitude within our society when it comes to the marijuana high. This view is deeply embedded into our modern Western attitude towards altered states of consciousness in general, as the American psychologist Charles Tart pointed out more than 40 years ago:

"Within Western culture we have strong negative attitudes toward altered states of mind: there is the normal (good) state of consciousness and there are pathological changes in consciousness. Most people make no further distinctions".[14]

The Many States of Consciousness

But are altered states of consciousness generally pathological? It should become obvious that there is something wrong with this view when we remember what kind of transformations our mind is going through almost daily. Every night we fall into an incredibly weird state of consciousness, experiencing illogical dreams with intricately interwoven storylines accompanied by vivid imagery. We are going through a phase that is somewhat like tripping on a strange magic mushroom yet this transformation occurs naturally, and repeatedly, every night. During sleep, we go through various states

---

[14] Tart, Charles (ed.) (1969/1990) *Altered States of Consciousness,* HarperCollins, San Francisco, p.2.

of REM (Rapid Eye Movement) and other phases in which our consciousness changes into something wild and mysterious.

"The Sleep of Reason Produces Monsters", painting by Francisco Goya, 1799

We also experience many types of altered states of mind while we are awake. When watching TV, surfing the Internet or playing a computer game we often fall into a weird state of immersion somewhere between trance and hypnosis. Many of us spend several hours every day in this altered state of mind immersed in virtual realities. We can go deeper into a state of trance listening or dancing to electronic music during a long club night. During our lifetime, we may typically go through thousands of prolonged phases of altered states of consciousness when we experience orgasms - intense ecstatic trips in which we usually deeply connect with somebody else during a strange dance performed by our bodies. We also go through other forms of ecstasy when we fall in love or sit in roller coasters, or when we party after winning an Ultimate Frisbee game or experience a runner's high during a marathon. We go through prolonged altered states of deep mental relaxation in a sauna, mineral bath or during a massage. When we get attacked, our mind automatically falls into a fight-or-flight or freeze mode, in which we become extremely alert, with our attention hyper focused on our opponent and possible survival strategies. In our modern society, we experience this fight-or-flight-or-freeze mode when we get into situations that lead to extreme stress, such as giving a presentation for the first time in front of a large audience. Many of us value the altered state of meditation – not only for relaxation but also for refreshment, for mental health and healing, to deal with various pressure, anxieties, grief, or traumas. We often go through phases of day-dreaming, in which we let our mind wander, less alert than usual and in a more imaginative state of mind.

We all value many of these altered states of consciousness and we even celebrate some of them. They can be highly functional and necessary for survival, like the concentrated tunnel vision in the fight-or-flight-or-freeze mode when we are physically attacked. Many such altered states of consciousness occur naturally and we usually understand how they can be useful in various ways. We embrace states like ecstasy or falling in love as essential and meaningful parts of our lives and usually believe that they are part of the essence of what makes our lives meaningful.

The Potential of Altered States of Consciousness

In some aspects, altered states of consciousness are superior to the rational mind state we usually consider to be normal. Brain imaging techniques for instance have already begun to answer Da Vinci's question why we can sometimes see things more clearly in our dreams, showing us hyper-stimulation in areas of the brain responsible for vision during our REM sleep phases.[15] These findings would explain why many creative, artistic people or scientists often report how they make use of their dreams.

In short then, we all naturally go through various transformations of consciousness every day and, obviously, we can profit from these changes enormously. Our consciousness has the natural ability to transform, and while some altered states of consciousness like hallucinating during a strong fever may be pathological, others are clearly evolutionary adaptive and helpful in our everyday lives.

However, we might ask, isn't it unnatural to induce an altered state of consciousness artificially by taking a mind-altering substance like marijuana?

---

[15] Compare Hobbson, Allan J. (2001) *The Dream Drugstore. Chemically Altered States of Consciousness*, MIT Press. Cambridge/MA.

Animals and Altered States of Consciousness

For thousands of years, humans in all cultures have invented methods to alter their states of consciousness with rhythms, music, dancing, meditation techniques and the use of psychoactive substances such as cannabis, alcohol, psilopsybin mushrooms, ayahuasca, or ibogaine. In our modern Western society, we often tend to look at these practices as outdated rituals, but when we take a close look at our society today, we find hundreds of millions of people all over the globe using a multitude of techniques including the use of psychoactive substances to transform their consciousness. Millions use music and substances like MDMA to arrive at an ecstatic trance state, others use the refined techniques of yoga to come to a state of deep meditation, others sing and chant together intoxicated at huge, traditional beer festivals.

An informed evolutionary perspective shows us that animals of all kinds systematically consume psychoactive substances to alter their consciousness. Some butterflies get drunk sipping on the alcohol of fermented fruits, cats get sexually aroused on catnip, goats eat coffee berries and frantically play around rolling down a hill. The psychopharmacologist Ronald K. Siegel intensively studied animals and their use of psychoactive plants for many years and concluded:

Various mushroom stones (approx. 1ft tall, 1000 B.C. to 500 A.D.

"In every country, in almost every class of animal, I found examples of not only the accidental but the intentional use of drugs. The thousands of cases I investigated convinced me that the action of an animal in seeking out intoxicants was a natural behavior in the animal kingdom."[16]

Siegel thinks that the search for intoxication is almost like a fourth drive – the three others being the drives for drink, food, and sex – and has an overall adaptive value for a species.

---

[16] Siegel, Ronald K. (1989, 2005) *Intoxication. The Universal Drive for Mind-Altering Substances*, Park Street Press, Vermont, p.13.

Altered states of consciousness belong to our existence and to a significant degree define who we are – and so does our curiosity and our ability to induce those states. They can be useful and substantially meaningful. Some of them are pathological, some are not. Obviously, even the altered states of consciousness that we find useful can also bring risks; even if some cognitive abilities may be enhanced, such as the ability for imagination during dreaming, others may decline, for instance our alertness to external goings on, which can certainly be dangerous in some situations. Dreaming may be a very useful state of consciousness, but you shouldn't do it while driving a car or operating a crane.

A New Perspective on The Marijuana High

A marijuana high can change and enhance many cognitive functions. Users have reported among other things a hyperfocus of attention, an enhanced ability to retrieve distant memories, to see patterns, to go through quick associative chains of thinking, to come to introspective and other insights, and to better empathically understand others. Other cognitive functions can decline during a high; your perception of time may be distorted and your ability to multitask often becomes worse, which can lead to considerable dangers in certain situations.

Those who want to positively use a marijuana high have to learn *how* to use it and, importantly, how to integrate the high into the rest of their existence – just like we have to learn how to integrate many other altered states of consciousness defining us. Just like sleeping and dreaming while driving, a state of trance or ecstasy will not be very helpful when operating dangerous machinery. But this does not mean that sleeping, dreaming, trance or ecstatic states are useless and only dangerous. Sleeping and dreaming are necessary for our survival. They are essentially a part of who we are. A routine like taking notes of your dreams right after you wake up, for instance, can help you to more meaningfully include your dreams in your life, to use them for creative purposes or to get to know your subconscious fears and desires. Taking short notes of your insights during a high can do a similar thing for you. As trivial as it sounds, this routine alone may significantly change what the high can do for you - and taking notes of ideas is just one of many measures that can help a marijuana user to better include the altered state of being high into his life.

Naturally, like so many others, you can just sit down and relax while you are high or enjoy the typical intensification of various sensations. You need to learn what strain works to relax you, how much you need to consume to get a certain effect and which environment is best for you. If you want to explore and positively use the more sophisticated effects of marijuana like enhanced pattern recognition, however, you have to learn more about how much to consume exactly for which activity, under which conditions and in which mood and condition. Like a cigar or wine aficionado, the high aficionado has a more profound knowledge about various strains and their chemical profile including details about growing and harvesting marijuana. So called 'Cannaficionados' or aficionados of cannabis, will also know the differences between highs coming off a joint, a bong, a pipe, a cookie, or from a vaporizer. Unlike the tobacco aficionado, however, marijuana users should not be mainly concerned with the taste experience but with the diverse effects potentially coming from various plants with different cannabinoid profiles through various routes of administration.

If we want to experience the marijuana high as an enrichment for our lives we first have to change our mind about our mind. We must first accept that we are far more than merely a rational, logically thinking being and give up the distorted self-image professed in our Western culture about the nature of our own consciousness. Only an altered perspective on altered states of consciousness will lead to the possibility of using the full potential of an altered state like the marijuana high – positively enriching individual lives and society in general.

Marijuana, Dopes, and Cognitive Enhancements

When genius mathematician and originator of cybernetics Norbert Wiener moved with his family from Cambridge to Newton, his wife organized the move and let Wiener concentrate on his work as a professor at MIT. She knew that her notoriously absent-minded husband would be of no help to her. She also knew he would forget that they had moved so she gave him a piece of paper with the new address of their home. Later during his work, Wiener spontaneously came up with an insight, found the piece of paper in his pocket, scribbled down the idea, but then finding an error in his workings, threw the piece of paper away. In the evening he drove home to his old address and soon realized that he did not live there anymore. The note in his pocket was gone. Wiener had no clue anymore where he lived, so he asked a little girl on the street: "Excuse me, perhaps you know me. I'm Norbert Wiener and we've just moved. Would you know where we've moved to?" The young girl replied, "Yes, Daddy. Mommy thought you would forget."

The Astrologer who Fell Into A Well, Illustration by John Tenniel 1884

The "absent-minded professor"-phenomenon has long become a stereotype and a popular character in countless comedies. We basically know what happens when an academic genius like Wiener shows this kind of absent mindedness. Wiener's attention was probably often so bound to thinking about current mathematical or other ideas or problems that there was not much 'processing power' left for navigating in the real world. We can assume then, that Wiener's 'dopey' behavior resulted from an intense process of concentrated thinking. Wiener's absent-minded behavior does not show that he is stupid; it only shows that some of his cognitive abilities – like his *spatio-temporal* orientation – suffered because some other cognitive processes were extremely active and enhanced.

We don't have to look at geniuses to learn this lesson. When you are really focused on upcoming exams you might run out the house with your shirt buttoned up the wrong way or wearing an odd pair of socks. The intensification of one cognitive process usually leads to weakening of others. We all know this from our everyday lives; it seems almost trivial to point this out, but we tend to forget this when it comes to the discussion of psychoactive substances and their effects on cognition. And as trivial as the lesson seems, it is an important one for users and scientists alike, as I will explain now.

Cognitive Enhancements

For thousands of years, marijuana users have reported in detail various forms of what we would now call *cognitive enhancements.* Some have used the marijuana high to hyperfocus their attention, to better attend to perceptions, memories, thoughts, or imaginations. Other users have observed that their minds are racing during a high, they feel they can think faster, associatively flying through thoughts, imaginations, or memories. Many value the marijuana high for retrieving distant memories from their past, episodic memories which often come in such astonishing detail that they feel as if they are being transported into the past. We have numerous reports of the enhancement of pattern recognition, where marijuana users describe how they suddenly discover new patterns during a high; patterns in a taste experience, in the behavior of other people, in art, in music, or in nature. Many users have also reported that they become more creative, that their ability for introspection and empathic understanding becomes enhanced or that they can generate great insights during a high.

As we all know, most of these cognitive enhancements can only be experienced given the right "set and setting". Also, the dosage must be appropriate and the user has to know how to "ride a high", which is an ability that involves theoretical knowledge as well as some skills. Importantly, this knowledge on the side of the user will involve knowing for which activity marijuana can be best used. This latter point becomes more obvious when we look at the downside of the enhancement of some cognitive abilities.

The Downsides

Of course, there are discussions even between users if marijuana can really enhance our cognition temporarily in those ways. In my book *High. Insights on Marijuana*[17], I have tried to show that under favorable conditions a marijuana high can in fact lead to all those enhancements. One thing seems obvious, though: these enhancements are never a complete enhancement of *all* of our cognitive abilities. Just as in real life, the intensification or enhancement of one cognitive ability usually comes at the cost of the worsening or slowing down of others. Here are three examples for some cognitive enhancements during a high and their downsides.

First, 'mind racing' during a high allows a cannabis user to quickly travel through memories, imaginations or chains of thought. Certainly, this enhancement can be used for many purposes but in my view, it might also contribute to the infamous short-term memory disruptions. So far, the following is only a hypothesis, but I believe that mind racing might actually be one of the reasons why we are loosing the thread during a conversation when we are high. In sports photography, if you take many photos in a quick succession, your camera's computer chip shuts down for some moments at a certain point – it needs to take a break to process and store the large amounts of data. Likewise, fast associative thinking might lead to working memory disruptions, where you may actually forget at some stage what the point of your story was.

Second, hyperfocusing during a high can intensify whatever comes within the focus of your attention and allows, for example, an enhanced analytic

[17] Marincolo, Sebastián (2010), *High. Insights on Marijuana.* Dog Ear Publishing, Indiana.

perception and understanding of the complex structure of a Miles Davies trumpet solo. On the other hand, the reinforcement of what cognitive scientists call "selective attention" during a high might make you act "dopey" like Norbert Wiener, causing you to phase out real world navigation tasks.

Third, many marijuana users have observed that during a high their ability to perceive patterns is enhanced. Some see new patterns in art and suddenly understand the influence of the work of Vincent Van Gogh on the style of painter Alexej von Jawlensky or better understand a pattern of shyness in the behavior of a friend. However, a marijuana high might lead to pattern over-interpretation, where we "see" faces in mountain tops or clouds. This may be useful for many purposes, but it might sometimes lead to misinterpretations of what is really out there.

Alexej von Jawlensky painting Alexander Sacharov, 1909

The enhancement of pattern recognition is a "double-edged sword", as Charles Tart observes:

"The patterns that are formed from visual data are organized into a degree of complexity and familiarity that is optimal for surviving in the world around us. Detecting a potential predator concealed in some bushes has survival value; seeing a potential predator in every ambiguous visual input is not conductive to survival of the organism. Thus we may conceive of some optical level (...) of pattern-making activity, of organization of ambiguous (and not so ambiguous) visual data into meaningful concepts. Raise this level too high and we have illusion and hallucination. Lower this level too much and we have stupidity. Marijuana seems to raise this level a fair amount, more so with increasing levels of intoxication."[18]

---

[18] Tart, Charles T. (1971), *On Being Stoned: A Psychological Study of Marijuana*

In brief, then, we can see that cognitive enhancements coming from a marijuana high usually lead to the weakening of other cognitive processes, just as we might expect it from our everyday lives. As trivial as this insight might seem, it has important implications both for marijuana users as well as for scientists.

Lessons to be learned

If used under favorable conditions, a marijuana high can lead to significant enhancements of specific cognitive functions and abilities. But these usually bring with them the weakening of specific other functions. Users have to learn how various strains can bring respective characteristic cognitive enhancements but also a temporary slowdown or other negative effects upon other cognitive functions. They have to learn for which specific activities to use a certain strain so that they can get the best out of the overall cognitive alterations deriving from its high.

There is a lesson to be learned from scientists working in the field, too. They must be careful with the implications of their studies when it comes to the evaluation of the effects of marijuana on cognition. If you only look for negative effects on cognition in specific situations (like, for example, performance in multitasking, various standard short-term memory tests, etc.), you will certainly find nothing else. We know that after decades of irrational prohibition, scientists were encouraged to obtain such conclusions. That's what they usually get paid for. Some of these tests may reflect that some cognitive functions during a high are disrupted or weakened in various situations. Yet, that does not imply that the reports of hundreds of marijuana users and their positive cognitive enhancements are wrong. Norbert Wiener might have been habitually bad at orientation, but that does not contradict the fact that the highly concentrated thinking and the resulting ideas causing his "silly" behavior were pretty stunning

---

Intoxication. Palo Alto, Cal.: Science and Behavior Books, p.59.

GUINEA – A Guerrilla-Neurophilosophical Approach to High Science

What is it like to be high? What is the positive potential of this altered state of consciousness for medical and inspirational uses? How can marijuana enhance cognitive functions like our ability to remember episodes, to recognize patterns, to introspect bodily states, to empathically understand others, and to generate insights? Can we systematically research the high? And how important is research in this field?

Cannabis flower photo © Sebastián Marincolo 2013

The Positive Impact of the High on Individuals and Cultures Worldwide

According to U.N. estimates, cannabis is the most widely cultivated and used psychoactive substance worldwide. In recent years, it has become more and more evident that medical marijuana has a huge therapeutic potential for numerous medical indications. Also, many use cannabis as a cognitive enhancer for inspirational and other purposes. Without doubt, we can also see a lot of abuse. Even though remarkably non-toxic, marijuana can certainly negatively affect the lives of consumers in various ways if abused. But many skilled users worldwide have discovered that a marijuana high can be an incredibly useful tool for many of their activities. We have innumerable reports from many cultures throughout history about how marijuana has helped individuals to work better creatively, to perceive all kinds of new patterns, to inspire them in their sexual relations, to come to introspective and other insights, to deepen their relationships with others and to personally grow. These enhancements have not only touched and transformed individual lives, through some individual marijuana users and their inspiring work, whole societies worldwide have profited from these positive effects.[19]

A concerted research effort of the marijuana high and its positive uses would give us a better picture of how cannabis use has actually influenced science, literature, art and other cultural developments worldwide. Innumerable inspirational and medical users would profit from knowing more about the exact cognitive alterations resulting from various strains. Moreover, researching the marijuana high may lead to important advances in our scientific understanding of the nature of mental processes such as introspection, empathic understanding, or the mostly unconscious process of generating insights.

In recent years we have learned that marijuana strains with high ratios of CBD (*cannabidiol*) can deliver an effective treatment for many medical indications without getting patients too high, which is a blessing. However, we should not turn away too quickly from the "high" as something we consider solely as an unwanted side effect. Many users still seek a certain high and its respective mind-alterations; the high will always remain crucial for certain therapeutic, inspirational, and other uses.

---

[19] Compare Ellen Komp's http://www.veryimportantpotheads.com/ for a list of influential marijuana users and a description of their respective uses.

Raphael Mechoulam, Prof. of Medicinal Chemistry at the Hebrew University of Jerusalem

The 'Gold Standard' for Medical and Inspirational Use

The experience of medical marijuana patients so far has clearly shown us that natural strains of cannabis work a lot better than synthetic cannabinoid extracts or natural extracts of single cannabinoids like THC. I therefore fully agree with medical expert Lester Grinspoon, who calls natural marijuana the "gold standard" for medical marijuana.[20] We have to bear in mind, then, that the various "highs" with their individual character valued by millions of users comes from natural strains with different phytochemical profiles. Cannabis strains contain more than 100 different cannabinoids, around 200 different terpenes and more than 20 flavonoids. Therefore, we should obviously expect significant and interesting variations in the high coming from different strains – the Israelian cannabis scientists S. Ben-Shabat and Raphael Mechoulam have called this the "entourage effect" of cannabis.[21] But how can we make serious progress when it comes to researching the ways in which various cannabis strains alter consciousness?

A New Approach to Researching the Marijuana High

In short, I would like to suggest GUINEA, a neurophilosophical approach to researching the marijuana high. GUINEA is an unconventional integrative and interdisciplinary approach that comes in two phases. First, we need careful anecdotal data mining of reports, which will mostly come from "guerrilla" users of marijuana. In other words, we need to critically analyze detailed reports of experienced marijuana users about their experiences with marijuana strains and each strain's mind-altering potential. This

---

[20] For an excellent essay on this subject see Malmo-Levine, David (2009) „Patented Pot versus The Herbal Gold Standard", http://www.cannabisculture.com/node/19879

[21] Ben-Shabat, Shimon (July 17, 1998). "An entourage effect: inactive endogenous fatty acid glycerol esters enhance 2-arachidonoyl-glycerol cannabinoid activity". *European Journal of Pharmacology* 353 (1): 23–31. doi:10.1016/S0014-2999(98)00392-6.

approach is not new. Both Harvard scholars Charles Tart and Lester Grinspoon started it more than 40 years ago.[22] While Tart sent a questionnaire about marijuana and its effects to 750 students and evaluated the feedback, Grinspoon collected reports from various literary and other resources. Tart's and Grinspoon's approach was based on the insight that set and setting significantly influence the high.

Dr. Lester Grinspoon, Associate Prof. Emeritus for Psychiatry at Harvard Medical School

Tart himself pointed out that the usual scientific method of observing subjects with no previous experience with marijuana in a laboratory setting just leads to anxiety and other reactions – reactions not caused by marijuana intake but by the sterile scientific setting. Another important fact which Lester Grinspoon and I have repeatedly stressed in the past is that much depends upon the *skill* of users to learn how to use marijuana and how to "ride a high" - just like surfers learn to use a surfboard to ride a wave.[23]

Ideally then, scientists should analyze stories of skilled *psychonauts* if they want to find out about the psychoactive potential of a high. Traditionally, scientists would work with inexperienced test persons when researching the effects of marijuana but, of course, they lack the knowledge to make use of the altered states of mind. More experienced users not only know how to use their strains for various enhancements given their own psychological make up and mood, they will also be able to introspect subtle nuances in cognitive changes brought about by various strains. A true wine connoisseur needs vast knowledge and a range of technical vocabulary about the nature of various grapes, soils, production and processing methods, as well as the storage of wines. The more he knows, the more he will be able to discriminate the nuances in his taste experience. For marijuana *psychonauts*, this kind of expert

---

[22] Tart, Charles T. (1971). *On Being Stoned: A Psychological Study of Marijuana Intoxication.* Palo Alto, Cal.: Science and Behavior Books, and Grinspoon, Lester (1971). *Marihuana Reconsidered.* Cambridge, M.A. Harvard University Press.

[23] Compare my essay http://sensiseeds.com/en/blog/marijuana-surfing-and-riding-a-marijuana-high/ A more detailed version is in my Marincolo, Sebastian *(2013) High. Das positive Potential von Marijuana,* Klett-Cotta/Tropen, Stuttgart.

knowledge will be even more crucial. For them, the issue will not be mainly about taste experiences, but about a whole array of cognitive enhancements. The better their knowledge of psychology, the more they will be able to introspectively tell us in which ways a strain leads to various cognitive alterations in such complicated processes such as attention Hyperfocusing, pattern recognition, creativity, or empathic understanding. Lester Grinspoon's Internet project "marijuana-uses.com", which collects reports and essays from experienced users is a unique resource for this approach because it presents a fine selection of reports from skilled marijuana users with great *psychonautic* abilities.

As a result of the critical evaluation of skilled user reports, we will be able to make hypotheses about the effects of marijuana strains on higher cognition. The formulation of those hypotheses involves not only experiential reports, but will include state of the art knowledge of various cognitive science disciplines. In the second phase of GUINEA, these hypotheses should be taken to empirical testing. In my previous book *High. Insights on Marijuana*, I have generated various hypotheses based on the analysis of anecdotal reports from skilled marijuana users which should be further taken to scientific testing, including hypotheses about possible ways in which a marijuana high might lead to certain changes and enhancements in bodily perception, introspection, empathic understanding, pattern recognition, and to an enhanced ability to generate insights.[24]

GUINEA is not only an interdisciplinary scientific approach, but also a cross-field one, meaning that we not only need to include knowledge from scientific disciplines such as the philosophy of mind, the neurosciences, psychology and other cognitive sciences, but also from other cultural and subcultural fields, such as the legal and Guerrilla cannabis growing scene or from shamanic and other cultures with experience of cannabis. Let us first look at the interdisciplinary scientific fields of knowledge involved.

---

24 Marincolo, Sebastián (2010), *High. Insights on Marijuana.* Dog Ear Publishing, Indiana.

Neurophilosophy and the High

If we want to understand how a marijuana high affects higher cognitive processes will have to turn to neurophilosophy - a modern interdisciplinary approach of various mind sciences. For instance, let's look at the question of whether a high can enhance empathic understanding. How does empathic understanding work, in general? In the last 20 years, there were heated debates in the philosophy of the mind and in cognitive and other fields of psychology. After the groundbreaking work by philosophers like Robert Gordon and Alvin Goldman, scientists focused their attention on neuroscientists who claimed to have found a specialized neuronal system - the mirror neuron system - underlying our capacity to feel with others and to understand them. The debate concerning empathy has led to fundamental advances in our understanding of the nature of empathy, which is extremely valuable when it comes to the understanding of pathological states such as autism or psychopathic behaviors.

Of course, many questions remain debated in this field. What is the exact role of the mirror neuron system? Autistic children have strong deficiencies in their empathic understanding; are these related to an underdeveloped mirror neuron system? Which cognitive abilities are involved in empathic understanding? It is important to see that these debates are still ongoing; the study of empathy today is a common interdisciplinary effort in which philosophers of the mind, neuroscientists, cognitive scientists, evolutionary biologists, developmental psychologists and many others are discussing new findings to integrate them into a unified picture of empathy. Likewise, whole arrays of researchers from various areas discuss other mental abilities, such as introspection or spontaneous insights. We have a myriad of reports that a marijuana high affects many of these complex mental abilities. Consequently, researching the high demands a similar interdisciplinary approach.

The Endocannabinoid System (ECS)

The most promising approach to researching the high within the neurosciences was certainly the discovery of one of the major physiological systems in our brain and body, the endocannabinoid system in the beginning of the

1990s. Interestingly, this system is found not only in humans but is also common to other vertebrate animals and can even be found in some invertebrate animals, which shows its importance in the process of evolution. As we now know, the effects of cannabis on both body and mind come because the cannabinoids from the plant are similar to those produced endogenously by our own brains. So, we all have a signaling system that involves endocannabinoids (anandamide and 2-AG) and endocannabinoid receptors (CB-1 and CB-2) involved in the regulation of various processes. The endocannabinoid system (ECS) is involved in the regulation of many bodily and mental processes including stress response, appetite, immune function, pain, sleep, and memory.

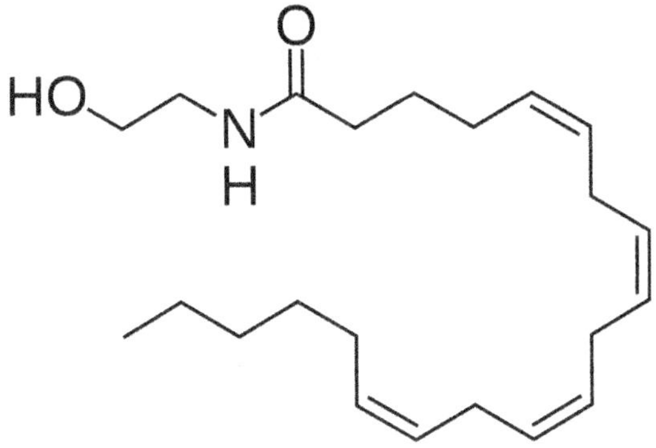

Skeletal formula of the endocannabinoid anandamide

New findings concerning the ECS in the last 20 years have to some degree explained many anecdotes about positive medical marijuana uses and inspirational uses. To name one well-known example, we now know that the endocannabinoid system also controls appetite; when consumed phytocannabinoids (from the cannabis plant) interfere with this signaling system, it may be stimulated to give you the "munchies", an effect that has been consistently reported by marijuana users for centuries. I will not go into the details of our knowledge of the ECS here. Let me just point out that although we already know about many of the functions of the incredibly versatile and important ECS in our brain, I believe that we are still at the beginning when it comes to understanding its role in various higher cognitive processes. Scientists investigating the ECS should definitely look into the details of marijuana user reports about the mind-altering potential as a guide to find more functions of the ECS in cognition.

Various Brain Imaging Techniques

Once we have generated scientific hypothesis based on critically evaluated anecdotal reports, we can start to design experiments with volunteers being subjected to brain scanning during a high. New imaging techniques like functional magnetic resonance imaging (fMRI), computer tomography (CT) or Positron Emission Tomography (PET) could augment our understanding of how the high affects various cognitive abilities such as pattern recognition, creativity, empathy, or introspection. Such practices could actually inform the neurosciences about possible functions of the endocannabinoid system and its functional relations to other neuronal systems. Obviously, we first need to have experiential reports of marijuana users about the cognitive effects of a high to know what we are looking for and to start constructing experiments.

Other Fields of Knowledge

The "neurophilosophical approach" to the study of the marijuana high will include just as many scientific disciplines as the general mind sciences studied today - evolutionary biology and psychology, a whole range of cognitive science disciplines, linguistics but also, for instance, an ethnobotanological perspective as can be found in the work of Ronald Siegel, with his groundbreaking research on the interaction between animals and psychoactive plants.[25] Also, scientists need

The American Robin has been observed to seek out intoxication from berries called Christmas holly. Ronald Siegel has shown that a wide range of animal species systematically seeks intoxication from plants in their environment.

[25] See Siegel, Ronald K. (1989, 2005), *Intoxication. The Universal Drive for Mind-Altering Substances. Park Street Press, Rochester Vermont.*

to understand that there is much relevant knowledge coming from various cannabis cultures and subcultures that have evolved over thousands of years. Scientists researching the marijuana high should be interested in *guerrilla* marijuana breeder's cannabis wisdom, shamanic knowledge, as well as looking at what professional agricultural cannabis engineers have to say about their marijuana strains. Without this background knowledge, we cannot really evaluate anecdotal reports connected to using those various strains and their effect on users. This is why I emphasize that the approach to researching the marijuana high should not only be scientifically "inter-disciplinary," but also "cross-field". Shamanic wisdom, the knowledge of cannabis breeders and anecdotal reports of skilled 'guerrilla' cannabis users, such as those collected by Grinspoon are usually not considered to be anything that should even enter the realms of science; but I believe they should actually be taken very seriously. A careful analysis of these fields of knowledge would help to lead the way for other scientists to come up with new hypothesis and experiments to investigate the nature of a cannabis high.

Philosophers, Neuroscientists, and Shamans

So far, there are no significant budgets for a concerted research program of the marijuana high. Governments worldwide have been heavily influenced by various interest groups to hinder the research of one of the most promising plants known to man. Presumably, an independently operating foundation would be a better place to look for the necessary open-mindedness and empathy for the millions of people who would profit from an interdisciplinary research center using GUINEA. It would be an unconventional decision to investigate the *positive* potential of the high – which is already outrageous to many. Supporters would also have to understand that our path of wisdom into the natural and fascinating jungle of the altered human mind cannot be an *autobahn* made of concrete that we brutally build into this jungle. Many empirical studies with inexperienced study participants taking a psychoactive substance for the first time have been so destructively intrusive in their whole setup that they show us more about the participants' reactions to their environment than about the effects of the respective psychoactive substance. The guerrilla-neurophilosophical approach of GUINEA would acknowledge the value of wisdom, not only of experienced guerrilla marijuana users but also of guerrilla breeders, shamans and other cannabis subcultures worldwide.

Importantly, this research project as a whole will not only help us to better understand what is going on when we get high and to understand the medical and inspirational potential of cannabis. It will also help us to find out more about the nature of human consciousness itself. Readers of Oliver Sacks' books should be well familiar with this approach. Sack's detailed anecdotes about his patients with neurological disorders or syndromes such as Tourette's syndrome do not only tell us much about the nature of respective pathological conditions, they help us to come to a new understanding of the nature of the human mind itself.[26]

The systematic changes of various cognitive processes such as, for instance, our empathic understanding during a marijuana high could be extremely interesting for researchers in their respective fields, leading to a better understanding of the functional architecture of those states themselves.[27] This is an important lesson: a cross-field neurophilosophical study of the marijuana high (or altered states in general) will not only help us to understand the altered state of mind of the high better, but also help us to understand the nature of the most complex abilities of the human mind, such as introspection, empathy, creativity and the ability to generate spontaneous insights.

To conclude, then, here is my recommendation for how to research the marijuana high: philosophers, cognitive scientists, psychologists, biologists, anthropologists, neuroscientists and others, research the high. Don't be shy. Work together. And don't be snobs. Guerrilla users, urban guerrilla growers, tribal shamans, or Indian sadhus may know a lot more than you imagine.

---

[26] Compare for instance Sacks, Oliver (1986). *The Man Who Mistook his Wife for a Hat.* London: Picador.

[27] See my essay "Marijuana, Empathy, and Cases of Severe Autism".

Part II Tune In

The Zen-Effect of Marijuana

"When walking, walk. When eating, eat."
Zen proverb

One of the most predominant effects of marijuana is a hyperfocus of attention during the high. Your attention seems to be strongly focused on whatever you choose to attend to. Often, this hyperfocus is experienced together with intensified or increased sensations. We know this effect from our everyday lives. If you eat in the dark and in silence, you can focus your attention better on the taste of the food without being distracted. As a result, your taste experience will be more intense – and you are able to better discriminate fine nuances in the taste experience. While you are high, you can likewise selectively attend to many more dimensions within these experiences and better explore subtle nuances. We have many reports from experienced marijuana users observing this hyperfocusing on all kinds of sensations, including auditory, visual, tactile, olfactory, taste or bodily sensations. In his seminal book "The Hasheesh Eater," the American author and *psychonaut* Fitz Hugh Ludlow notes:

Fitz Hugh Ludlow, 1836-1870

"(h)asheesh always brings with it an wakening of perception which magnifies the smallest sensation till it occupies immense boundaries."

Another marijuana user describing the stronger focus of attention during a high expressed what he likes about marijuana as follows:

"I like all sorts of things about marijuana. (...) I like the first rush when time turns slow and liquid, and my hearing goes acute, and my focus intensifies. I like the way food tastes, the way country air smells, and the way water

goes down my throat cool and clean. (...) I like the way a fine single malt scotch tastes when there is a nice coating of herb on my tongue, like velvet fire."[28]

Hyperfocusing during a Marijuana High

Generally, the hyperfocusing of attention during a marijuana high seems to lead to an intensification of sensory experiences and often to a better analytic capacity for introspectively exploring these states and processes. In addition, it can also lead to a pronounced focus on thoughts, feelings, moods, imaginations, or memories.

In his amazing article "The Effects of Marijuana on Consciousness", psychology professor Arthur C. Hastings compared what I call a "hyperfocus" effect on attention with "psychological tunnel vision":

"The process of attention is clearly affected by marijuana. The most obvious effect is to narrow the amount of diverse contents in the focus of attention. The person under marijuana usually perceives fewer objects of attention, which may mean physical objects, actions, social elements, emotions, etc. We have already noted this effect: a person who is high may become absorbed in an object, event, or process to the exclusion of everything else. A train of fantasy may occupy all of a person's attention. This is a psychological analogy to tunnel vision, with the contents of the tunnel expanded."[29]

Hyperfocusing and Zen Buddhism

With their minds unusually bound and absorbed by sensations, thoughts or fantasies, many users have noted that they have a strong feeling of being

---

[28] Cross, Mackkenzie (2010) "What I like about Marijuana", in: Grinspoon, Lester (2014) (ed.), marijuana-uses.com

[29] Hastings, Arthur C. (1969) "The Effects of Marijuana on Consciousness." In Charles Tart (1969/1990), *Altered States of Consciousness*, HarperCollins, San Francisco, Third Edition 1990, p.407-32. Hasting's incredibly insightful article is a must-read for anybody interested in the psychological marijuana high.

in the "here and now" – or, if focused on their memories intensely reliving a moment of the past, being in the "here and then". They generally cease to share their attention between various tasks or stimuli – they do not eat a sandwich while talking to somebody on the phone and watching television. During a high, users tend to dwell on the immediacy of the moment, fully and joyfully appreciating whatever comes to their attention. Clearly, this is one of the teachings of Zen Buddhism: to focus on one activity and to appreciate the magical sensation of the moment, to be in the here and now. I therefore sometimes call this effect of marijuana the "Zen-Effect". Many marijuana lovers use this effect to enrich their lives. They use it to relax and be in the here and now when sitting on a dune watching the sunset, to deeply explore the experience of watching a waterfall or a painting by Edgar Degas, to better enjoy intensified tactile sensations during sex, or to simply concentrate on the way legendary double bass player Scott LaFaro unfolds his full potential with the Bill Evans Trio.

Use and Abuse of the Zen-Effect

The Zen-effect of marijuana holds amazing potential for users, especially for a modern society which demands us to constantly share our attention, jumping from telephone and mobile calls to writing emails, engaging in chat-rooms, attending meetings, or reading notes on business network portals. Simultaneously, we are permanently distracted by advertisements lurking around every corner of our real and virtual existences. A marijuana high can be positively used to actually sit down and concentrate, to breathe, to be in the here and now, to come back to your senses, to intensely experience your own current bodily feelings, but also to concentrate on one's memories, to enjoy a movie without dis-

Buddha statue, Thailand. Photo © Sebastián Marincolo 2011

tractions or to simply appreciate the fact that you are alive and well, sitting next to your beloved partner.

However, obviously, a marijuana high can be abused to constantly focus on the here and now in order to screen out problems in life. The adolescent marijuana abuser who permanently gets high to flee from problems with parents, teachers, and others is a commonly found example of abuse. Where marijuana can actually have a negative impact that should not be underestimated. Sadly, prohibition has led to an enormous lack of objective education about marijuana, including the vital distinction between use and abuse. Many people deprived of this information had negative experiences with marijuana in their younger years and then refrained from marijuana completely and never explored the positive potential of the plant. A great number of these past abusers therefore remain sceptical regarding this potential.

The Zen-effect is a good case to illustrate that the effects of marijuana can only be profitable if users learn and understand how to use the marijuana high and productively integrate it into their lives. Scientific investigations in the last few decades have almost exclusively been driven by a prohibitionist attitude based on misconceptions resulting from a long-term disinformation campaign against marijuana. Certainly, it is interesting to understand the risks of marijuana consumption, yet such studies were almost exclusively designed to show for which activities a marijuana high may not really be helpful, or even detrimental.

Re-focusing on the Potential of a Marijuana High

We as a society need to free ourselves from the old myths concerning marijuana, myths that have been installed by a decades-long disinformation campaign. This propaganda has locked our focus on marijuana risks - risks that to a large degree have been invented by spin-doctors. We need to shift our attention. We have to start investigating the potential of marijuana - not only for industrial use or for various "physical" medical uses, but also the potential of the marijuana high – for inspirational, recreational, as well as psychological medical uses. Of course, we should not exclusively hyperfocus on the positive potential – we still want to know more about possible risks and therefore be able to educate the public about these risks – but it is now time to take a deeper look and explore the positive potential of the high as an altered state of consciousness. So far, the prohibitionists are still making the important decisions concerning the course of scientific research. They should be reminded of the words of the Chinese philosopher *Lao Tsu* in his famous book *Tao Teh King*, a dictum which holds true not only for the subject of marijuana:

"If something exists which cannot be fully revealed to him with his viewpoint, he does not demand of it that it be nothing but what it seems to him."

Marijuana and the Enhancement of Episodic Memory

"A memory is a beautiful thing, it's almost a desire that you miss."
Gustave Flaubert, French writer (1821-1880)

Our memory is simply amazing. We can store information about thousands of situations but also abstract knowledge, theories, learned routines as well as practical skills like ice-skating or driving a car. Our memory keeps relevant information poised to meet our daily needs and plans, enables us to recognize people and places and facilitates remembering past events. With its various functions, memory constitutes the cognitive basis for our personal identity, for our decision-making, our present actions, and our attitude towards the future. Our memory defines who we are. Memory is everything.

Short-Term Memory Disruptions

When it comes to marijuana and its acute effect on memory during a high, many people think only of the infamous disruptions of our short-term (or "working space") memory during a high. The effect is nicely described in Jack Margolis' and Richard Chlorfene's book "A Child's Garden of Grass" in a short conversation of two very stoned consumers:

Virginia: Are you hungry?
Andy: No. (Long reflective pause). Wait a minute. Did you mean am I hungry for food, or am I hungry in the abstract, like hungry for knowledge or adventure?
Virginia: What were we talking about?
Andy: You asked if I were hungry.
Virginia: Did I?
Andy: Yes.
Virginia: Well, are you?
Andy: Am I what?

When using various kinds of marijuana it is fascinating to see how much this effect depends not only on the dosage used but also on the type of marijuana and the route of administration. A freshly harvested strain with higher levels of the terpene pinene, consumed with a vaporizer may bring almost no such disruptions and users often enjoy its high for hours of discussion or other activities without experiencing short-term memory problems.

Enhancement of Episodic Memory Retrieval

On the other hand, many marijuana users have reported an enhancement of what cognitive scientists have called "episodic memory" or "autobiographical memory". From these reports we know that during a high, users can often retrieve long-forgotten episodes of their lives, or have a much more vivid recollection of past events than usual. According to Yawger, the philosopher John Stuart Mill (1806 – 1873) mentioned the memory enhancing effect of cannabis:

"John Stuart Mill ... wrote of (cannabis's) power to revive forgotten memories, and in my inquiries, smokers have frequently informed me that while under the influence, they are able to recall things long forgotten."[30]

In the last few decades, scientists have almost completely ignored these reports, presumably because government funds were given mainly to research only the negative effects of marijuana. Yet, there are so many detailed reports about enhanced episodic memories that it seems outrageous to scientifically ignore this phenomenon for so long. On Lester Grinspoon's website marijuana-uses.com, a 19-year old computer programmer "Mackenzie Cross" writes in his piece "What I like about Marijuana":

"Memories seemed to force themselves upon me, very rapid but very gentle. I started to remember things in my childhood that made me truly happy and joyful. Things I had either forgotten or just simply didn't give the time of day to. I remembered raising my hands up as a signal for my mother that I wanted to be carried and the utter joy I felt when she would reach down and pull me up to her chest. I realized how much she really did, in fact, love me when I remembered how I longed for her goodnight kisses, of which never ran dry."

---

[30] Yawger, N.S. (1938). "Marijuana: Our New Addiction." *American Med. Sci.*,195, p. 353.

Likewise, Jeremy Wells, a 24-year-old undergraduate student of history, tells us about the recollection of childhood memories during a high in his report "Four Leaf Clovers", featured on the same website:

"Recently I was visiting with a relative who has a two-year-old baby girl, and we were looking for four-leaf clovers. So here I was, a 24-year-old man on my hands and knees, combing through the grass and screaming, "over here I found one". Actually we found several and I think we were probably more excited than the kid. See children have a way of doing that to you. Through them you can vicariously relive your childhood. In the name of "playing with the kids" you can shed your inhibitions and do the things you used to take for granted. You look at the world through a different set of eyes. It alters your worldview. In the course of reflecting on my four-leaf clover (of course, I kept one) I began to think of all the things you used to do as a kid, and how everything holds wonder and magic for children."

Episodic Memory, Introspection, and Empathy

These two reports also show that the more vivid retrieval of a past episode in one's life during a marijuana high often brings back not just the episode as such, but also our personal stance and feelings at the time. Enhanced episodic memory during a high could be at least a partial explanation as to why many marijuana users report introspective, as well as empathic, insights into others. If you have better access to the person you used to be in various phases of your life, you will have a better understanding of how you became who you are now – which aspects of you have changed, where you have made progress, and where you have not. And if you remember better how you felt as a kid, you will have more understanding of children and their needs, fears and hopes, from their perspective.

Our episodic memory is crucial for introspection and empathy, but also for many other kinds of cognitive activities. The phenomenon of enhanced episodic memory during a marijuana high has been reported so often that it definitely deserves a closer look from psychologists and cognitive scientists.

We now know that our episodic memory relies on activities in the frontal cortex and the hippocampus. Both these regions of the brain contain cannabinoid receptors that belong to an endocannabinoid signaling system and also react to the consumption of THC. So far, however, we do not know if there is a direct way in which marijuana consumption enhances episodic memory retrieval and recollection, or if this is an indirect effect. An indirect route could be that a high usually leads to a hyperfocus of attention – this is what I believe to be one of the most basic effects of marijuana – which may lead to a more vivid and intense recollection of an episodic memory on which we focus.

I think it should be obvious that research concerning marijuana and the often observed enhancement of episodic memory will not only pay off for those interested in marijuana, but could actually deliver highly relevant facts about the nature of episodic memory itself, including the possible involvement of the endocannabinoid system.

Flaubert's Reminder

The French writer Charles Baudelaire wrote his book *"The Artificial Paradises"* based on his experiences with marijuana and other substances. When he sent his manuscript to his friend Gustave Flaubert (author of "Madame Bovary") in 1860, Baudelaire received a very warm and admiring response about the many brilliant observations in it. Flaubert had also participated in some meetings in Paris of the "Club of the Hasheesh Eaters" ("Club des Hashishins"), of which Baudelaire was a member. Beside his overall praise for Baudelaire's observations, Flaubert remarked:

French writer Gustave Flaubert, 1821-1880

"It seems to me that concerning this subject field, (...) in a work which is the beginning of a science, in a work of scientific observation and inductive reasoning, you emphasize too much (at various places) the spirit of evil. One can feel here and there the sour influence of Catholicism."[31]

Maybe scientists and politicians today need to overcome the 'sour' influence of the flawed thinking behind marijuana prohibition in order to invest in research that could help so many of us to better understand its temporary effects on episodic memory.

---

[31] Gustave Flaubert in a letter to Charles Baudelaire in 1860, in: Müller, Ulf, and Zöllner, Michael, (ed.) (2002)*Der Haschisch Club. Ein literarischer Drogentrip*, Tropen Verlag Stuttgart.

Marijuana and The Power of Imagination

"The imagination is the golden path to everywhere"
Terence McKenna, philosopher, psychonaut, ethnobotanist, 1946-2000

"The true sign of intelligence is not knowledge but imagination."
Albert Einstein, 1879-1955

The Underestimated Value of Imagination

When we think about our capacity for imagination, we often think of it as a kind of visual daydreaming – a handy ability utilized mainly by artists and other creative people. We tend to underestimate how crucial our use of imagination is in our everyday lives. In fact, we rely on our imagination all of the time, not only when we think creatively, but also when we make decisions. You may decide not to go skiing next week because you heard the bad weather forecast and imagine it to be very cold and nasty in the mountains. Anticipating the cold wind up there in the mountains, you can literally feel a shiver going down your spine. As this example shows you not only visually "picture" situations in your mind when you use your imagination; you also imagine sounds, tastes, smells, feelings or moods. In his seminal book *Human. The Science Behind What Makes Us Unique*, neuroscientist Michael Gazzaniga reminds us how powerful a role imagination plays in our lives:

"Imagination also allows us to time travel. We can go in the future and back

to the past. An event may be long in the past, but I can replay it in my imagination from memory. (...) Likewise, I can project in the future. I can use my past experience of an emotion and apply it to future circumstances. I can imagine how I would feel, for example standing at the open aircraft door with a parachute on my back (terror, which I have felt in the past and did not enjoy) and decide I can bypass this adventure."

Of course, there are more important decisions in life than those about whether we want to go skiing or jump out of a plane. When we make decisions about whether or not to marry our beloved partner, we might go through several processes of imagination: "Can I see us getting along also in times of crisis? Do I enjoy the thought of having her by my side as the mother of our children? Will she still make me a better person in 20 years?"

It is easy to see how the capacity for imagination leads to an evolutionary advantage, as the neuroscientist Vilayanur Ramachandran observes:

"When you imagine something – as when you are rehearsing a forthcoming bison hunt or amorous encounter – many of the same brain circuits are activated as when you really do something. This allows you to practice scenarios in an internal simulation without incurring the energy cost or risk of a real rehearsal."

Imagination is not just a nice-to-have ability mostly developed in creative professionals. We all rely heavily on our imaginative capabilities every day. Our imaginative abilities have a fundamental impact on the way we all make decisions in our lives.

The Enhancement of Imagination During a High

It is important to be reminded of the importance of imagination to our cognition to understand the importance of the enhancement of imagination during a marijuana high. Many users of marijuana have reported a more vivid imagination when reading or thinking about a situation. When Harvard professor Charles Tart sent out questionnaires for his psychological study *On Being Stoned,* many users of marijuana endorsed the following statements for moderate levels of a marijuana high:

"If I try to visualize something, I see it in my mind's eye more intensely, more sharply than when straight.' (...)
"If I try to have an auditory image, hear something in my mind, remember a sound, it is more vivid than when straight." (...)
"If I try to imagine what something tastes like, I can do so very vividly."
"I can experience vivid tactual imagery, imagine what things feel like and feel their texture very vividly in my mind"

As a concrete example, here is a report from a mandolin player in a bluegrass group, who writes that he likes to practice playing high:

"I might smoke before playing. I play in a group, and I'll sit down and do a couple of hits to put a little edge on while I'm playing. When I'm stoned, I can visualize musical relationships more easily. The other day, I was practicing scales on the mandolin, double lines of scales in intervals. Playing them high, I made more sense out of them, and finally understood when and how they might be useful in my playing." [32]

Why don't We Connect the Dots?

Many marijuana users have experienced several enhancements of imagination themselves and have described how they use these enhancements. Some have reported that they can better visualize objects or faces and use this in their artwork; musicians feel that they can better imagine how various instruments work together and use this for musical compositions, whereas others like to get high and start cooking, using the enhancement of their imagination to come up with new ideas for recipes, imagining how certain spices or herbs would taste with certain vegetables.

I believe that most marijuana users have experienced the enhancement of imagination during a high in some form, but underestimate the incredible cognitive potential of this effect. Why don't they connect the dots?

---

[32] Novak, William (1980) *High Culture. Marijuana in the Lives of Americans*, The Cannabis Institute of America, Massachusetts, p.52.

There are several reasons why we tend to underestimate the potential of this imaginary enhancement. First, an ongoing disinformation campaign that has lasted for decades has led us to focus on the risks of marijuana instead of looking at its potential. Second, we tend to look at the process of imagination as a sort of conscious daydreaming and forget that we also rely on it *unconsciously* way more often in our everyday thinking and decision-making. Third, we usually consider imagination only as *visual* imagination; but as stated above, imagination is much more than that. Fourth, we tend to think more of other psychoactive substances like LSD or psilopsybin when it comes to enhanced imagery and imagination, because these substances bring full-blown visual trips.

Once we understand that marijuana has the potential to enhance our imagination, and also understand how important imagination is to our cognition in general, we can better understand the many reports of scientists, artists, writers and other people who have used this enhancement for a whole variety of purposes.

Scientists have often described how their imagination helped them come up with solutions to problems and great ideas. One of Einstein's greatest insights came when he imagined what it would be like to ride on a beam of light, while the chemist Friedrich August Kekulé von Stradonitz saw the benzene ring in a reverie of a snake biting its tail. The designer Philip Stark reported in an interview how he actively uses his dreams to better imagine and visualize new designs in 3D.

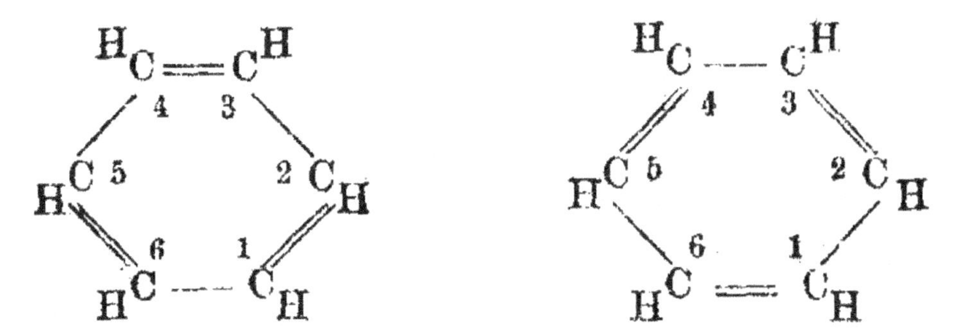

The chemical structure of benzene as proposed by Kekulé

Naturally, a marijuana high alone will not turn you into an Einstein, a Kekulé or a Philip Stark. But it may help you to better imagine situations, objects, smells, sounds, tastes or tactile sensations. If you are able to "ride your high" and to use this enhancement, it could help you with important decisions, to vividly remember crucial events in your life, or to come up with life changing ideas.

Enhanced Imagination and Politics

The German philosopher Ernst Bloch participated in hashish experiments with his friend Walter Benjamin in 1928 and later praised the value of enhanced imagination during a hashish high in his central work "*The Principle of Hope*", written between 1938 and 1947. Bloch compared the state of a cannabis high to that of a waking dream and emphasized that this altered state of mind allows us to think clearly and to profit from the experience:

The German philosopher Ernst Bloch, 1885-1977. Photo: Krueger, Wikipedia commons

"(...) opium seems to be more associated with the dream of the night, whereas hashish seems associated to the freely wandering, swarming daydream. Also, during hashish intoxication the ego is hardly altered, neither its individual nature nor its reasoning ability are taken away (…) Modern experimenters have characterized Hashish dreams as of an enchanting lightness (…), in short, for the talented hashish dreamer, the world becomes a request concert." (...) "Furthermore, hashish intoxication does not lack another kind of lightness: 'the individual believes to see his plans disentangled and to move towards their realization, plans which seemed impossible to clarify so far.[33]*"*[34]

---

[33] Lewin, Louis (1927), *Phantastica. Die betäubenden und erregenden Genußmittel – für Ärzte und Nichtärzte*, p.159 ff.

[34] Bloch, Ernst (1959), *Das Prinzip Hoffnung*, Suhrkamp Verlag, Frankfurt p.100.

Bloch, usually considered as a neo-Marxist thinker, emphasized that humans are far more than simply the product of their environment and upbringing; his philosophy is based on an image of human as essentially striving to think ahead and to imaginatively work on a utopia. He would later become a vastly influential professor for philosophy at the University of *Tübingen* – where I would later study and complete my doctorate in philosophy – and became friends with Rudi Dutschke, both of whom would later be considered as leading figures in the 1968 student revolution in Germany. Bloch clearly saw the positive social and political dimension of this state of daydreaming induced by hashish:

"Daydreams have no censure whatsoever from a moralistic Ego, as we can see it in the dreams of the night: their overreaching utopian Ego builds itself and its own as a cloud-castle in an astonishing unburdened blue. (...) The Self of the daydream can become so wide, that it also stands for others: (...) The sleeper is alone with his treasures; the Ego of the daydreamer can extend to others. If the Self (is) not introverted anymore or just related to its immediate environment, then its daydream seeks to improve public matters."[35]

Ernst Bloch's "The Principle of Hope" first appeared in 1954 and became a set text for many of the Beat Generation, decades before they started to daydream, inspired by the use of cannabis.

In 1972, marijuana user John Lennon, who had been turned on to cannabis by Bob Dylan in New York some years earlier, almost got deported back to the UK for having been caught with marijuana. Lester Grinspoon, who testified for Lennon during the legal process, recounts how he sat down for dinner with Lennon and Yoko Ono:

"I told John how cannabis helped me to hear his music for the first time in much the same way that Allen Ginsberg had seen Cezanne for the first time after he smoked cannabis before going to the Museum of Modern Art, to determine if marijuana could help him break through his incapacity to relate to Cezanne. It did.

---

[35] *Ibid.*, p.101.

John replied that I had experienced only one facet of what marijuana could do for music. He said that it also enhanced the ability for composing and making music."[36]

Shortly before the deportation hearings, Lennon had expressed his daydreams in his song "Imagine" – the song that would later become one of the hymns of the peace movement:

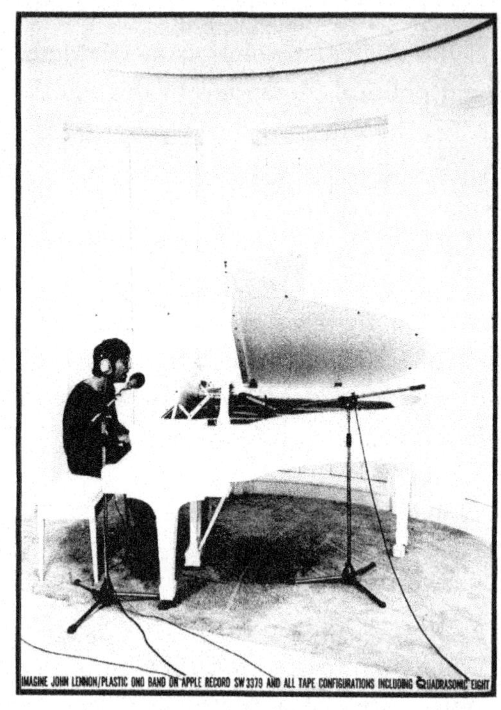

Imagine
there's no countries
It isn't hard to do
Nothing to kill or die for
And no religion, too

Imagine
all the people
Living life in peace

You, you may say I'm a dreamer
But I'm not the only one
I hope someday you will join us
And the world will be as one.

---

[36] "Brady, Pete" (2001), "Lester Grinspoon: Defending Medical Marijuana," *Cannabis Culture Magazine*, http://www.cannabisculture.com/articles/2013.html.

Marijuana and The Slowdown of Time Perception

"Oh! Do not attack me with your watch. A watch is always too fast or too slow. I cannot be dictated to by a watch."
Jane Austen, Mansfield Park, English novelist, 1775 - 1817

Hurled Away by a Stream of Ideas

The phenomenon of a slowed down perception of time during a high is one of the most well known effects of marijuana – infamous to some, highly valued by others. Of course, these distortions of time perception can be seen solely as a risk for users – and it is certainly true that those perceptual changes during a high can become dangerous while, for example, driving a car. On the other hand, many users appreciate this change of perception in safe situations as one of the most valuable experiences during a marijuana high. We have detailed reports about the subjective slowdown of time from members of the "Club des Hashischins" ("Club of the Hasheesh Eaters. The members of this cannabis club ingested large doses of hash marmalade, so it comes as no surprise that many of them became familiar with this phenomenon that manifests itself especially under stronger doses. Charles Baudelaire, one the founding members of the club, wrote:

French writer Charles Baudelaire, 1821-1867

"... a new stream of ideas carries you away: it will hurl you along in its living vortex for a further minute; and this minute, too, will be an eternity, for the normal relation between time and the individual has been completely upset by the multitude and intensity of sensations and ideas. You seem to live several men's lives in the space of an hour." [37]

[37] Baudelaire, Charles (1860) "Le Poème du Haschisch", in: *Artificial Paradises, Citadel, 1998.*

Baudelaire's statement already hints at two effects of marijuana that I believe to be relevant for the subjective effect of a slowdown of time perception. He describes something a friend of mine once called "*mind racing*" during a high: "a stream of ideas … will hurl you along … and the individual has been completely upset by the multitude (…) of sensations and ideas". Baudelaire further notes the "intensity" of sensations and ideas. In my view, this intensity of experience could also come from what I have called "hyperfocusing" during a high. When high, we hyperfocus on sensations, thoughts or our imaginations and often forget about what is going on around us. Whatever comes into focus becomes more intense. We know a similar but more subtle effect from our everyday experience: When you close your eyes and take time to focus on the taste of ice-cream melting in your mouth, your taste experience gets more intense; you also perceive more details. A forced focus of attention always brings more intensity to whatever we attend to.

Fitz Hugh Ludlow in Eternity

In his famous book "The Hasheesh Eater" (1857), the American author Fitz Hugh Ludlow gave us an even more detailed description of the effect of the perceptual slowdown of time during a strong high. Like Baudelaire, Ludlow also ingested large doses of hashish and was also completely absorbed by his intense imaginations and streams of thought during his high:

"The thought struck me that I would compare my time with other people's. I looked at my watch, found that its minute hand stood at the quarter mark past eleven, and, returning it to my pocket, abandoned myself to reflections. Presently, I saw myself a gnome imprisoned (…) in the Domdaniel caverns, "under the roots of the ocean". Here (…) was I doomed to hold the lamp that lit that abysmal darkness, while my heart, like a giant clock, ticked solemnly the remaining years of time. Now, this hallucination departing, I heard in the solitude of the night outside the sound of a wondrous heaving sea. Its waves in sublime cadence, rolled forward till they met the foundations of the building; (…) Now, through the street, with measured thread, an armed host passed by. The heavy beat of their footfall and the grinding of their brazen corslet rings alone broke the silence

(...). And now, in another life, I remembered the fact that far back in the cycles I had looked at my watch to measure the time which I passed. (...) The minute hand stood half way between fifteen and sixteen minutes past eleven. The watch must have stopped; I held it to my ear, no, it was still going. I had traveled through all that immeasurable chain of dreams in thirty seconds. "My god!" I cried, "I am in eternity."[38]

"Mind-Racing", Hyperfocusing, and Associative Leaps in Thinking

Like Baudelaire, we can see how Ludlow's mind is racing. He is going through so many associative chains of thought and detailed imaginations that it feels to him like a long time must have been passed since he began his reverie. Usually, he would need hours or days to go through those detailed reflections which actually only lasted for 30 seconds. Also, similar to Baudelaire, Ludlow describes that he is completely absorbed by his thoughts; in other words, he hyperfocuses on an inner stream of thought and, thus, does not pay attention to other processes around him that are unfolding in real time. Both accelerated thinking – or, as I like to call it, "mind-racing", as well as the hyperfocus of attention, then, are described in Ludlow's report of a radical slowdown of his perception of time.

Ludlow's story, however, adds another interesting aspect to Baudelaire's report: His illustrious associative "chain of dreams" is not only detailed and long, but it is also "jumpy": with various associative leaps, he jumps from one detailed imagined situation to another ("Now ... and now ... and now ..."). The unusual associative leaps, which are often reported about a high, probably add to his subjective feeling that he 'mind-travelled' a long distance, much longer than he would usually do in 30 seconds or even in 30 hours.

According to the reports above, the slowdown of time perception could then arise out of the effects of marijuana to lead to an attentional hyperfocus

---

[38] Ludlow, Fritz Hugh (1857/2009), *The Hasheesh Eater: Being Passages from the Life of a Pythagorean.* Chapter II: "Under the shadows of Esculapius." http://www.lycaeum.org/nepenthes/Ludlow/THE/index.html.

of the perceiver on an unusually accelerated stream of thoughts, a stream characterized by 'jumpy' associative thoughts or imaginations.

It is of course possible that marijuana has an even more direct neurological effect on the way we perceive time. Neuroscientists are beginning to better understand the neural substrates of interval timing, and it seems that these areas are abundant with cannabinoid receptors.[39] However, as far as I can see, we have so far no clear understanding on the exact involvement of the cannabinoids in time perception.

The Uses of Time Slowdown

This account of the perceptual slowdown of time during a high would explain why many users of marijuana appreciate the effect of a perceptual time slowdown so much. With their racing and concentrated mind, users often find themselves to be better able to appreciate the subtleties and depths of immediate sensations. For them, the subjective slowdown of time is not merely a perceptual distortion, but can become a real mind enhancement. Arguably, the slowdown of time perception during a high played a fundamental role in the evolution of jazz music and in the evolution of other musical genres such as reggae. I'll discuss the role of marijuana in the early evolution of jazz later in my essay “Vipers, Muggles, and The Evolution of Jazz.”

Joyful experiences like eating a great meal, looking at a beautiful piece of art, or skiing down a mountain seem wonderfully prolonged. In his magnificent essay “Mr. X” featured in Lester Grinspoon's book “Marijuana Reconsidered”, an anonymous author wrote:

“The actual duration of orgasm seems to lengthen greatly, but this may be the usual experience of time expansion which comes with cannabis smoking.”[40]

---

[39] Atakan *et al.* (2012), “The effects of cannabis on perception of time: a critical review.” *Curr. Pharm. Des.*, 18(32): pp. 4915-22.

[40] Grinspoon, Lester (1971), *Marijuana Reconsidered*, Harvard University Press, Harvard, p.126.

Lester Grinspoon would reveal the identity of the author after his untimely death; it was his best friend Carl Sagan.

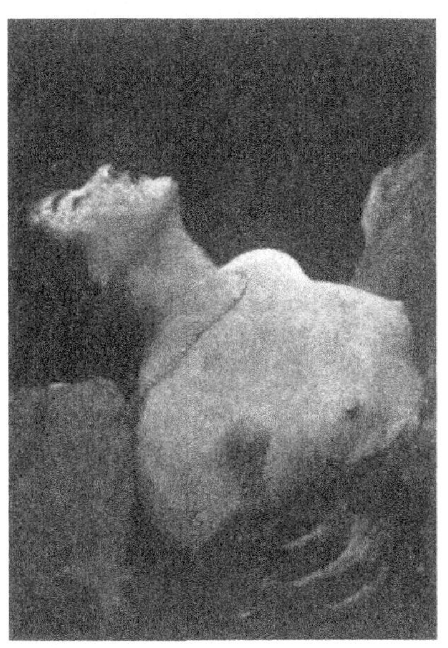

"*Sinnenrausch*", painting by Zmurko Franciszek (1859-1910), around 1890

The slowdown of time perception can be used for relaxation, to step outside the ever-accelerating speed of the routines and demands of our modern everyday lives, and to simply enjoy being in the here and now of existence. Countless users seek this perceptual slowdown to appreciate seemingly endless moments in which they can taste incredible details and nuances in a great wine, hear the complex and soothing sounds of gentle waves washing upon a beach at night or to go on a seemingly infinite voyage during lovemaking. And, as the psychoanalytic Erich Fromm once said:

"Who will tell whether one happy moment of love or the joy of breathing or walking on a bright morning and smelling the fresh air is not worth all the suffering and effort which life implies."

Marijuana, Pattern Recognition, and What it Means to be 'High'

"To understand is to perceive patterns."
Isaiah Berlin, philosopher, 1909-1997

"As long as habit and routine dictate the pattern of living, new dimensions of the soul will not emerge."
Henry van Dyke, short story writer and essayist, 1852-1933

The Importance of Pattern Recognition

When we talk about patterns or pattern recognition, we tend to think of simple visual patterns like a striped blanket or a dotted curtain. But our pattern recognition abilities are way more sophisticated than just recognizing basic visual designs like that. We can visually recognize and distinguish types of trees, cars, or the different painting styles of particular artists. And we perceive not only visual patterns in our environment, but also hear patterns in sounds or music; we perceive the tactile pattern of a wooden surface, the gustatory pattern of the taste of a mango and we can intellectually recognize patterns such as a certain defensive tactic style used by an opponent in a chess game. We easily recognize thousands of human faces despite their changes over time, and recognize places and objects even under completely different lighting conditions. We can learn to recognize the sound pattern of a humming bird, the roaring of a lion or the inimitable style of legendary pianist Dinu Lipatti interpreting a Chopin waltz. Patterns are everywhere, and their perception, memory and recognition are imperative to our survival.

Enhanced Pattern Recognition During a Marijuana High

We have numerous reports from marijuana consumers about the enhancements of various types of pattern recognition abilities. The Harvard psychology professor Charles Tart evaluated enhanced pattern recognition as a *characteristic* effect observed by marijuana users during a high in his study "On Being Stoned". Approximately, one third of the questioned students agreed with the following statement as correct for a strong or very strong high:

"I can see patterns, forms, figures, meaningful designs in visual material that does not have any particular form when I'm straight, that is just a meaningless series of shapes or lines when I'm straight"[41]

Tart's study shows that marijuana users not only report this enhanced pattern recognition ability for visual patterns, but also for the perception of all kinds of patterns perceived through various sensory channels. I myself have repeatedly noted that during a high I find myself able to better understand foreign languages, a fact also noted by some contributors to Lester Grinspoon's website project marijuana-uses.com (see "Lady Chatterley Stoned" and "Some experiences With Language and Learning"). Another anonymous contributor "Rob" describes how a high made him recognize a pattern of "timid rigidity" in his walking style:

"I was also fortunate in that I got high that night, as many reefer virgins report not getting high the first time they smoke. An amazing understanding came to me while walking home. As I strolled along the tree-lined sidewalk carrying on a conversation with a friend, I felt an awkward stiffness in my stride, realizing with each step a timid rigidity. I have since altered my manner of walking to a confident, open gait of long strides and silent footsteps."

Rob's experience shows what many reports about enhanced pattern recognition during a high have in common - a shift in attention. Rob does not primarily attend to the conversation he was having with his friend any-

---

[41] Tart, Charles T. (1971), *On Being Stoned: A Psychological Study of Marijuana Intoxication.* Palo Alto, Cal.: Science and Behavior Books, p.71.

more, but to his own body and his walking style, allowing him to perceive a pattern he was unaware of before. Other marijuana users have reported similar pattern recognition processes, where they shifted their attention away from the verbal content of a conversation to the body language of their conversation partner.

Why do we Discover New Patterns During a High?

I am sure that there are many more alterations of cognitive processes during a marijuana high which allow users to better perceive patterns of which they were previously unaware. The hyperfocusing of attention during a high certainly helps many users to come to a more acute perception of patterns in whatever they attend to. Also, an enhanced episodic memory probably helps to *'re-cognize'* patterns, which are similar to patterns seen before. If I better remember my last encounter with a Dalmatian dog, this will help me to better see the pattern of a Dalmatian in this picture.

On a deeper, neurological level, I have suggested in my previous book, *High. Insights on Marijuana*, that marijuana leads to a synesthetic effect (or, with lower doses, to a *'pre-synesthetic'* effect) that according to the neuroscientist Vilayanur Ramachandran plays a major role in the explanation of pattern recognition.

Rob's story also beautifully illustrates that the recognition of a pattern plays a major role in transcending this pattern. Rob had to first understand that his walking actually follows a pattern to be able to make a conscious decision to break it, because he felt that he wanted to change this pattern. Ultimately, then, a marijuana high not only allows a user to perceive new patterns in nature, music, art, in the workings of society, in the behavior of others or in himself. It also allows him to break with unwanted and previously unconscious patterns of thinking or behavior. This is not only one of the most important preconditions for creativity, but also an almost inexhaustible source for personal growth and can also be used to get over patterns of addiction. A marijuana high may lead you to understand that you follow a routine pattern in lovemaking and help you to transcend it. You may find that you follow a learned routine when dealing with family issues or that you keep on reacting to stress in a bad way. If you use marijuana in the right way, if you learn to surf a marijuana high, it can help you to understand and get away from many unhealthy routines and patterns of behavior, which is why it can be helpful in the treatment of various addictions and in helping you to grow as a person.

What it Means to be 'High

Marijuana can alter your consciousness in a way that enables you to take a different perspective on patterns in life, as if you were looking from high up, to see and evaluate your live and the lives of others in a privileged way. You can see different patterns in life and in reality, patterns unfolding in space and time, patterns which are normally invisible to you. It is as if you were standing on a hill looking down at your neighborhood and realizing the generic pattern in which you and your neighbors have spent your lives, in those “little boxes on the hillside, little boxes made of ticky-tacky, (...) little boxes all the same”, as Malvina Reynolds once expressed in what would later become the theme song to the acclaimed comedy series “Weeds”. Maybe this euphoric and elevating feeling of looking at life's patterns from this 'high' perspective, and then transcending them, is an inspiration behind the slang term 'high' itself.

The Effects of Marijuana on Body Image Perception

"There is more wisdom in your body than in your deepest philosophy."

Friedrich Nietzsche, philosopher, 1844-1900

How do we localize and identify an itch right under our shoulder? A burning feeling of pain in the neck? How does our brain manage to balance on two feet when we walk, and how do we know how our body is located in space even when we close our eyes? In their book *The Body Has a Mind of its Own*, Sandra and Matthew Blakeslee sum up their answer to these questions:

"Every point on your body, each internal organ and every point in space out to the end of your fingertips, is mapped inside your brain. Your ability to sense, move, and act in a the physical world arises from a rich network of flexible body maps distributed throughout the brain – maps that grow, shrink, and morph to suit their needs." [42]

The Body Mapping System and the Enhancement of Bodily Sensations

The psychologist Charles Tart mentions in his study "On Being Stoned" that marijuana users feel that a high can help them to better introspectively access their current bodily sensations and feelings. Many of the users surveyed confirmed the following effects as a "common" effect of a marijuana high:

"My skin feels exceptionally sensitive"
"Pain is more intense if I concentrate on it"
"My perception of how my body is shaped gets strange; the 'felt' shape or form does not correspond to its actual form (e.g. you may feel lopsided, or parts of your body feel heavy while others feel light"
"I feel a lot of pleasant warmth inside my body"

---

[42] Blakeslee, Sandra, and Blakeslee, Matthew (2007), *The Body has a Mind of Its Own.* Random House Trade Paperbacks, New York, p.7.

"I am much more aware of the beating of my heart"
"I become aware of breathing and can feel the breath flowing in and out of my throat as well as filling my lungs"[43]

According to neuroscientist Antonio Damasio, we all have an *interoceptive* sense which gives us a sense of the body's interior. For Damasio, this inner sense rests on a representational mapping system, also called the *somatosensory* system, which has developed in order for us to screen internal bodily states such as *"(...) pain states, body temperature, flush, itch, tickle, shudder, visceral and genital sensations; (...)."*[44]

Could it be that marijuana interacts with this body mapping system? We also know that large doses of marijuana can lead to "body image distortions", which are also often described in literature. In a report of a stoned experienced under the influence of a high dose of ingested hashish, the American writer Bayard Taylor wrote as far back as 1863:

Bayard Taylor, 1825-78.

"The sense of limitation – of the confinement of our senses within the bounds of our own flesh and blood – instantly fell away. The walls of my frame were burst outward and tumbled into ruin; and, without thinking what form I wore – loosing sight even of all idea of form – I felt that I existed throughout a vast extend of space. The blood, pulsed from my heart, sped through uncounted leagues before it reached my extremities; the air drawn into my lungs expanded into seas of limpid ether, and the arch of my skull was broader than the vault of heaven."[45]

---

43 Tart, Charles T. (1971). *On Being Stoned: A Psychological Study of Marijuana Intoxication.* Palo Alto, Cal.: Science and Behavior Books, pp. 109-126.

44 Damasio, Antonio (2003). *Looking for Spinoza. Joy, Sorrow, and the Feeling Brain.* Harcourt Books., p.106.

45 Taylor, Bayard (1863). *The Land of The Saracens, or Pictures of Palestine*, Chapter X, Asia Minor, Sicily and Spain. http://www.online-literature.com/bayard-taylor/lands-of-the-saracen/(hypertext version 2009).

This report shows that a marijuana high can strongly affect ones body imaging system. It also shows that it is a matter of dosage and use whether or not marijuana can help with introspection. Whereas a decent high might help with intensifying bodily sensations and bringing them to our awareness, a higher dose might lead to misrepresentations and distortions of our inner body mapping system.

How to Lose Your Body During a High

Tart also notes in his study that under stronger doses of marijuana, users often seem to forget about their bodies. Many users agreed with the following descriptions as common effects for a strong high:

"I lose awareness of most of my body unless I specifically focus my attention there, or some particularly strong stimulus demands my attention there."

"I have lost all consciousness of my body during fantasy trips, i.e. gotten so absorbed in what was going on in my head that my body might as well not have existed for a while."

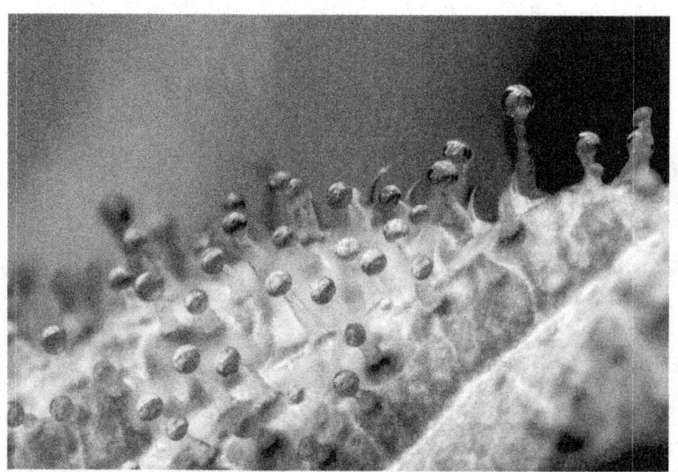

The mushroom-shaped trichomes, where the cannabis plant builds its cannabinoids. Photo © Sebastián Marincolo 2012.

"I have lost all consciousness of my body and the external world, and I just found myself floating in limitless space (not necessarily physical space)."[46]

How can we explain that? One explanation could be that even

[46] Tart, Charles T. (1971). *On Being Stoned: A Psychological Study of Marijuana Intoxication.* Palo Alto, Cal.: Science and Behavior Books, p.110.

though marijuana stimulates the body imaging system, it also stimulates other capacities such as imagination and episodic memories in such a way that our attention sometimes gets strongly drawn away from our bodily sensations and shifts to fantasies or memories.

Ultimately, though, this is just a hypothesis at this point.

Endocannabinoids, Exogenous Cannabinoids, and the Body Imaging System

These observations raise interesting questions. How does marijuana affect the body mapping system? We know that other effects of marijuana strongly depend on its interaction with an already existing endocannabinoid system in our brain. We also know, for instance, that endocannabinoids play a role when it comes to the process of thermoregulation; but what are the other functions in the body mapping system? How do various endocannabinoids play a role in the regulation of this system? And how do exogenous cannabinoids – that is, cannabinoids produced externally by the consumption of cannabis – play a role in affecting this system? Will a strain with higher levels of cannabidiol (CBD) affect the body imaging system in a different way to other strains? I am convinced that these questions might lead to really interesting developments in science and medicine. It is high time to investigate.

Marijuana, Empathy, and Severe Cases of Autism

"You can only understand people if you feel them in yourself."
John Steinbeck, writer, 1902-1968

"Could a greater miracle take place than for us to look through each other's eye for an instant?"
Henry David Thoreau, writer, 1817-1862

Empathy and the Simulation Theory

I have always been fascinated to see in how many ways a marijuana high can enhance empathic skills in users. My interest in this effect of marijuana was first sparked when I experienced it myself about 15 years ago. I noticed it with excitement because I had already had a long-standing philosophical interest and research focus on theories of empathic understanding.

In contemporary philosophy of mind, philosophers like Robert Gordon, Alvin Goldman and my teacher Simon Blackburn at the University of North Carolina at Chapel Hill took a fresh look at theories of human understanding and empathy in the late 1980s, arguing for a version of what would later become known as the "simulation theory of human understanding". Before the development of the simulation theory, cognitive scientists and philosophers had thought that we understand humans on the basis of a "folk psychology", a quasi-theoretical body of psychological knowledge that allows us to make generalizations and give explanations how people feel and behave. The claim was that we would all learn this folk-psychology as we grew up by learning to use psychological vocabulary to describe, predict, and explain the behavior of others. This position was later labeled as the "theory-theory" because the theory basically relied on the claim that we all – mostly unconsciously – use something like a psychological theory when understanding others.

Briefly, the simulation theory stated that in order to understand others, we use a special cognitive ability to "put ourselves in the shoes of other people". In other words, rather than just using a psychological theory about others, we understand them by simulating them, looking at the world from their point of view. Looking for an empirical confirmation, proponents of the simulation theory argued that many autistics (especially high-functioning autistics) would be able to grasp theoretical psychological concepts and generalizations, but would have deficiencies when imaginatively simulating others, which would explain their problems with empathic understanding. For the past decades, cases of Autistic Spectrum Disorder (ASD) have remained in the minds of philosophers, psychologists & cognitive and neuroscientists when it comes to theories of human understanding and empathy.

Marijuana and the Enhancement of Empathic Understanding

During research for my first marijuana study for *High. Insights on Marijuana,*[47] I found many astonishing reports from users about various enhancements of their empathic skills during a high. A busy father described how he got high before he played with his son and for the first time understood how alone his boy feels and how much he craved for more of his father's time and attention. A husband wrote a letter to his wife explaining to her how the marijuana high enabled him to better understand her needs during sex. A psychotherapist reported that he always talked to his patients in a sober state of mind, but one day got high in private and then received an emergency call from a patient. His patient was so impressed with his empathic skills during the conversation that she later insisted on paying for the hour. These and other reports prompted me to think about possible explanations for the enhancement of our fundamental skills to simulate and understand other people during a high.[48]

Many of the cognitive enhancements during a high could play a role in the enhancement of empathic skills. Countless marijuana users have observed and described enhancements like an enhanced episodic memory or an enhanced ability to recognize patterns during a high. Such enhanced

---

[47] Sebastián Marincolo (2010), *High. Insights on Marijuana.* Indiana: Dog Ear Publishing

[48] For more reports compare Lester Grinspoon (2014), *marijuana-uses.com*, and Novak, William (1980). *High Culture: Marijuana in the Lives of Americans.* Massachusetts: The Cannabis Institute of America, Inc.

cognitive abilities can obviously help with empathic understanding; if I can vividly remember episodes of my teenage years, it stands to reason that I will be able to better understand a teenager experiencing similar situations. If I can better recognize the subtle pattern of a sarcastic smile in the face of my conversation partner, I can better understand how that person feels and acts towards me. Importantly, besides these and some other possibly relevant cognitive enhancements, many of the reports of marijuana users explicitly stated that a high can help them to 'slip into another person', to feel his feelings, to see his point of view. In an intriguing report, Théophile Gautier, another member of the famous 19th century literary circle 'Club des Hashischins, describes this imaginative perspectival change during a high while looking at a painting:

Théophile Gautier (1811-1872) in a portrait of Auguste de Chatillon, 1839

"By some bizarre prodigy, after several minutes of contemplation I would melt into the object looked at, and I myself would become that object. Thus I turned into a nymph Syrinx, since the fresco represented Leda's daughter pursued by Pan. I felt all the terrors of the poor fugitive, and sought to hide behind the fantastic reeds to avoid the ram-footed monster."[49]

Reports of this kind made it obvious to me that marijuana can fundamentally enhance our ability to simulate others and to take on their point of view.

[49] Gautier, Théophile, (1966). "The Hashish Club." In: Solomon (ed.) (1966), *The Marihuana Papers*, Signet Books, Indiana, p.174.

The Simulation Theory and the Mirror Neuron System

The debate concerning the simulation theory took on a new twist when an Italian group of researchers around Giacomo Rizzolatti discovered the mirror neuron system in the early 1990s. In short, the group noticed that when a monkey grabbed a peanut, the same group of motor neurons responsible for its hand movement would fire not only during the grabbing, but also when the monkey merely perceived someone else grabbing the peanut. Since this finding, neuroscientists like Rizzolatti, Vilayanur Ramachandran, and Marco Iacoboni have argued that mirror neurons comprise a specialized system of neurons that fundamentally subserve our ability to "mirror" and to understand the emotions and intentions of other people. Simulation theorists used this line of research to argue for their position: a specialized mirror neuron system would actually constitute our special capacity to simulate others when we understand them "from inside", rather than just making folk-psychological inferences about them.

The Italian neurophysiologist Giacomo Rizzolatti

Marijuana, Autism, and the Endocannabinoid System

In 2006, Vilayanur Ramachandran published a paper entitled "Broken Mirrors – A Theory of Autism"[50], arguing that autism could be linked to a defective ("broken") mirror neuron system - a highly controversial theory which is still subject to much debate now. Based on my own research, I introduced a hypothesis on a possible connection between the endocannabinoid system and the mirror neuron system in a chapter of my first book on marijuana and empathy:

[50] Vilayanur S. Ramachandran & Lindsay M. Oberman, (2006) "Broken Mirrors: A Theory of Autism", *Scientific American* 295, pp.62-69. doi:10.1038/scientificamerican, pp.1106-62.

"Could it be that (...) there already exists a functional relation between the endocannabinoid system in our brain and the body mapping system, including the mirror neuron system? Again, a look at the enhancements of cognitive skill under marijuana may be fruitful for a general scientific outlook on the workings of the human brain."[51]

Now, if there is such a functional connection, could it be that the endocannabinoid system is defective in autistic children, causing their problems with empathic understanding? Recent findings suggest that I was roughly on the right track, even though the "broken-mirror"-hypothesis remains highly controversial. In the following I will first describe how some severely autistic children seem to profit incredibly from medical marijuana and then summarize some new findings on possible links between the endocannabinoid system and autism.

Severe Autistic Spectrum Disorder and Medical Marijuana

When we are talking about cases of severe kinds of Autism Spectrum Disorder (ASD) it is important to emphasize that we are not discussing personalities like Big Bang Theory's Sheldon Cooper or the eponymous movie character, Rainman, played by Dustin Hoffmann. Children with severe forms of ASD are often highly auto-aggressive and/or aggressive towards other people; typically, they act up with intense tantrums, they don't engage in interactive, pretend or imaginative play and prefer solitary or ritualistic play. They have difficulties with verbal and non-verbal communication – some never learn to speak at all – and usually they don't make friends, are withdrawn, don't maintain to eye contact, show a lack of empathic understanding, are emotionally unstable and get angry or distressed when routines are changed. Autistic children often seem to be unhappy, almost in agony, acting angrily or aggressively as if in severe pain. Autism now affects 1 in 68 children in the US and prevalence figures are growing.

In the last six years, some courageous parents have overcome their prejudices against what their society taught them is an "evil drug" and listened to

[51] Sebastián Marincolo (2010) "High. Insights on Marijuana", Indiana: Dog Ear Publishing

some experts who advised them to try medical marijuana for their children. Many parents had felt that the pharmacological drugs prescribed by their doctors for ASD did not work or had even worsened the condition of their children. For some parents, turning to medical marijuana was definitely a dramatic decision, as they desperately tried to save the lives of their children on the verge of dying.

In the following, I will cite some of their observations about the treatment of their children with medical marijuana with a special focus on the enhancement of the child's abilities to socially engage with others. Several parents have observed that marijuana helps to calm their children, to make them smile and happy and to greatly diminish their tantrums and aggressive behaviors, but it is crucial to understand that we are not simply talking about an effect of sedation. On the contrary, many parents described that after medication with cannabis their children seemed to become more alive and joyful, were able to perform actions they had never previously been able to do, engaging in social interaction and curious to explore new activities. A father reports about the behavior of his eight-year-old autistic son Sam, 30 minutes after giving him a small dose of marijuana

"(H)is behavior became relaxed and far less anxious than he had been at the time we gave him the MC {medical cannabis}. He started laughing for the first time in weeks. My wife and I were astonished with the effect. It was as if all the anxiety, rage and hostility that had been haunting him melted away. That afternoon and evening his behavior was steady and calm. He started talking to us and interacting with us again. Sam was physically more relaxed and began initiating physical contact with the motivation being affection instead of aggression. It was amazing!"[52]

Mieke Hester Perez, the mother of a 12-year-old autistic son, reports:

"He was on a combination of thirteen prescription drugs, and his weight dropped down to 46 pounds. He was diagnosed with anorexia and malnutrition,

[52] Hester Perez, Mieke (2014), http://weedpress.wordpress.com/science/illnesses-marijuana-helps/autism/sam's-story-using-medical-cannabis-to-treat-autism-spectrum-disorder/

second to his autism. (...) Ultimately, his doctors gave him six months to live. I was devastated. And I was determined I would figure out a way to extend his life."

Then she gave him medical marijuana:

"The immediate change I saw was eye contact. He gained over 40 pounds, he's happier and better behaved."[53]

Some of these parents have no doubt that marijuana saved their children's lives. And while it may sound less dramatic in that context, Perez's report on eye contact is remarkable; maintained eye contact signifies the ability to socially interact with a person and to better understand facial expressions.

Severely autistic children have problems to mimic and learn behavior. C.B., the father of a nine-year-old severely autistic son told me how his son managed to put on his shirt and pullover himself for the first time in his life and started to curiously explore new activities after receiving his medical marijuana. About 30-45 min after ingestion of cannabis, the screaming, the agony, the emotional instability were mostly gone. His son smiles and laughs a lot and "just wants to cuddle". The positive change in his son's behavior after cannabis medication is simply breathtaking.

Another now prominent mother of an autistic son, Myung Ok-Lee, wrote about the effects of medical marijuana on her severely autistic son "J":

"The experts don't live in my house, nor do they get to reap the rewards, like this morning, when J woke up, smiled, and wanted a hug—the boy who formerly woke us with a scream of pain. The boy who, since he was 3 years old, never gave us hugs or let himself be hugged, because he couldn't bear to be touched. (Fittingly, the next person he bestowed a hug to was Organic Guy, his grower.) Now, when he's proud of something, like his awesome bike riding skills, he glances to find my face, to make sure I'm looking."[54]

---

[53] Zouves, Natasha (2012), http://www.neontommy.com/news/2012/01/ryan-s-story-medical-marijuana-and-autism.

[54] Ok Lee, Myung (2011) "Why I give my 9-year old son pot, Part 4. Two years in, and

Another flower of a cannabis plant. Photo © Sebastián Marincolo 2013

Then, Ok-Lee took her son to his doctor while he was under the influence of medical marijuana:

During the pre-exam discussion, J's pediatrician was both taken aback and a bit sceptical to hear about his new cannabis "therapy." But when I brought J into the examination room, she saw that he didn't look the least bit stoned, which had been her big fear. Instead, he said "Hi" to her and patiently (for him) allowed her do the exam, which was a first – usually he can't bear be to be touched, especially around his head. But this time, J even let her stick the tickly otoscope in his ears and shine a light in his eyes. He said "Ah" on command. Last time, she couldn't listen to his heart because he kept grabbing the stethoscope off her head. This time – after he had a listen first – he handed the stethoscope back

---

I'm still flying solo."
http://www.slate.com/articles/double_x/doublex/2011/05/why_i_give_my_autistic_son_pot_part_4.2.html.

to her and let her finish the exam."[55]

Marijuana and the Enhancement of Empathic Skills

Marijuana seems to help some autistic children in many ways: it makes them happier, calm, less anxious, less auto-aggressive. Aggressive behaviors get vastly diminished and compulsive behaviors are reduced, to name just a few. But what parents often describe as a miracle is the change in their social and empathic skills; after being medicated with marijuana, autistic children often initiate and keep eye contact, start to engage in social communication, start to play with other kids instead of attacking them, they enjoy to touch and to be touched. They give hugs and allow themselves to be hugged, and they often show mimicry behaviors that they were never capable of previously, like rinsing a bowl after eating or putting on a t-shirt. These observations are important in the context of empathy because this kind of mimicking – or, 'mirroring' – the actions of others is a basic capacity that rests on the ability to put oneself in the position of another person. For us, it seems so easy to imitate or "mirror" behavior that we do not even understand, yet for a severely autistic child, this imitation behavior is often not possible. They seem to lack the ability to imaginatively put themselves in the position of their mother when she rinses the bowl and then to repeat the action themselves. Likewise, they seem to have problems to use their imagination to put themselves in the position of a person in pain or in joy and to feel their feelings as if they were their own, as we often do. When we watch somebody breaking an arm in an accident, we suffer; when we watch a pair in love kissing passionately, we delve into happy memories of similar experiences. Severely autistic sufferers cannot recreate the feeling and, thus, do not often seem to understand the feelings of others.

Empathic understanding is probably the most complex and most outstanding mental skill we – and some animals – have and it is an incredibly powerful drive for our actions. More than 150 years ago, the German philosopher Arthur Schopenhauer wrote:

---

[55] *Ibid.*

"How is it possible that suffering that is neither my own nor of my concern should immediately affect me as though it were my own, and with such force that it moves me to action". ... This is something really mysterious, something for which reason can provide no explanation, and for which no basis can be found in practical experience. It is nevertheless a common occurrence, and everyone has the experience. It is not unknown even to the most hard-hearted and self-interested. Examples appear every day before our eyes of instant responses of the kind, without reflection, one person helping another, coming to his aid, even setting his own life in clear danger for someone whom he has seen for the first time, having nothing more in mind that that the other is in need and in peril of his life".[56]

How can cannabis so dramatically improve this skill in various ways – in autistic children as well as in other people, as reports of users suggest? Could it be that the endocannbinoid system is functionally involved in the very neurological systems that are subserving our empathic skills? As we will see, recent scientific findings indeed point in this direction.

Autistic Spectrum Disorder and the Endocannabinoid-System

For about 5000 years now, cannabis has been used worldwide to treat a whole range of diseases, syndromes and medical conditions, including nausea, neurological pain, epilepsy, glaucoma, epilepsy, Tourette's syndrome, asthma, inflammation and autoimmune diseases as well as countless other conditions.

The discovery of the endocannabinoid system (ECS) in the mid-1990s was not only a revolution in neuroscience, it also brought about a better understanding why and how exogenous (consumed) cannabis could have such a profound healing effect for so many conditions. The more we learn about the ECS, the more it becomes evident that it plays a crucial functional role in a multitude of bodily and mental processes. It is now known to be a cen-

[56] In: Arthur Schopenhauer, *Die beiden Grundprobleme der Ethik, II*, Über das Fundament der Moral, (1840) (*Sämtliche Werke*, Volume XII, Cotta'sche Buchhandlung, Stuttgart, (1895-1898).

tral component of health and healing processes in the body. As Pacher *et al.*, (2006) point out, there is a growing body of evidence showing that deficiencies of the ECS lead to various diseases.[57]

Recently, more scientific studies have suggested direct links between Autistic Spectrum Disorder (ASD) and the ECS. Let me briefly explain some of those links:

A study published by Chakrabarti and Baron-Cohen tested whether variations in the endocannabinoid receptor 1 (CNR1) could be associated with the duration of a human's gaze towards happy faces. Atypical gaze fixation patterns are typical for neurodevelopmental conditions like autism. The study rests on research in primates showing that the striatal region of the brain *"plays a major role in the directing gaze. The striatum is thought to encode a 'value map' of the visual stimuli."*[58]

The study further proceeds from the fact that the endocannabinoid system is one of the key systems involved in the functioning of the striatal circuit. In their conclusion, they state:

"These results suggest that CNR1 variations modulate the striatal function that underlies the perception of signals of social reward, such as happy faces. This suggests that CNR1 is a key element in the molecular architecture of perception of certain basic emotions. This may have implications for understanding neurodevelopmental conditions marked by atypical eye contact and facial emotion processing, such as ASC [autism spectrum conditions]."[59]

A study published by Stanford scientists Földy *et al.* 2013 in *Neuron*[60] concerns the protein neuroligin 3, post-synaptic cell adhesion molecules in-

[57] Pacher, Pál, Sánddor Bátkai, and George Kunos (2006), "The endocannabinoid system as an emerging target of pharmacotherapy." *Pharmacological review* 58.3.

[58] Chakrabarti, B., and Baron-Cohen, Simon, (2011) "Variation in the human cannabinoid receptor CNR1 gene modulates gaze duration for happy faces", *Molecular Autism* 2011, 2:10, p.11.

[59] *Ibid.*, p.10.

[60] Földy, C., Malenka, RC, Südhof, TC. (2013) "Autism-associated neuroligin-3 mutations commonly disrupt tonic endocannabinoid signaling." *Neuron.* May 8;78(3):

volved in the communication between brain cells. Rare mutations in these molecules are know to predispose to the syndrome of autism and are also involved in a certain type of secretion of endocannabinoids in the brain: Földy *et al.* summarize their results as follows:

"Our data thus suggests that neuroligin-3 is specifically required for tonic endo-cannabinoid signaling, raising the possibility that alterations in endocannabinoid signaling may contribute to autism pathophysiology."[61]

Another link between the ECS and ASD is based on the observation that the cannabinoid receptor CB2 is upregulated in those with ASD as well as in other neurodegenerative disorders.[62] The upregulation of CB2 receptors as a response to damages support the thesis that the ECS serves as an endogenous neuroprotective system and that consumed (exogenous) cannabinoids activating these CB2 receptors are promising therapeutic agents. Benito et al. suggest that the upregulation of CB2 receptors in those conditions might not only be an endogenous neuroprotective response, but could also point to a possible role of CB2's in their pathogenesis.[63]

Empathy, Cannabis and the Endocannabinoid System

These findings suggest a possible link between deficits in the endocannabinoid system and autistic spectrum disorder. Could it be that the endocannabinoid system is involved in subserving some cognitive functions underlying our ability to empathically understand others? I have introduced the "broken mirror-theory" of ASD before, which states that it is caused by deficiencies in the mirror neuron system. There is still a lot of controversy over the issue of the function of the mirror neuron system itself and whether it plays a role in ASD. Critics argue that the mirror neuron system does not

---

pp.498-509. doi: 10.1016/*j.neuron*.2013.02.036. Epub 2013, April 11.

61 *Ibid.*, p.498.

62 See Siniscalco *et al.*(2013) *"Cannabinoid Receptor Type 2, but not Type 1, is Up-Regulated in Peripheral Blood Mononuclear Cells of Children Affected by Autistic Disorders"*, *J Autism* Dev Disord DOI 10.1007/s10803-013-1824-9

63 Benito, C. *et al.* (2008) "Cannabinoid CB2 receptors in human brain inflammation." *British Journal of Pharmacology* 153.2: pp.277-285.

play a crucial role for empathic understanding and that the empirical evidence for its implication in ASD is sparse.[64]

As those discussions are going on, I believe we should direct our attention to the obvious connection between cannabis, the endocannabinoid system and empathic understanding. I have quoted some of the great amount of anecdotal evidence for an enhancement of empathic understanding during a high. I then also reported how incredibly helpful cannabis can be for severely autistic children to relate empathically to others. Obviously, something fundamental is going on here. We should look for possible functional links between the endocannabinoid system and cognitive functions subserving empathic understanding. As mentioned above, we have already evidence that endocannabinoids are involved in the perception of basic emotions. As far as I know, nobody has so far found a connection between the mirror neuron system and the endocannabinoid system, but I think it is possible that we might find substantial links here – links that could be revolutionary in our understanding of the human mind in general, and pathological conditions like ASD in particular. I would recommend for other scientists in this interdisciplinary field to take a closer look at detailed reports of marijuana users and at what parents have to say about the astonishing effects of medical marijuana on their children with ASD.

'Mexican Sativa' strain flowering. Photo © Sebastián Marincolo

[64] For a good overview of the debate see Thomas, Ben (2012), "What's so special about Mirror Neurons", http://blogs.scientificamerican.com/guest-blog/2012/11/06/whats-so-special-about-mirror-neurons/

Severe Cases of Autism: Dying Patients Can't Wait

While interested observers and scientists can calmly sit back and do more research, severely autistic children and their parents cannot wait. Some of these children are dying because they cannot eat, are banging their heads on walls or sharp objects every day or are suffering the horrible side effects of several pharmaceutical products which do not help them at all or often even worsen their condition. They are screaming in agony, locked in their tantrums and shaken by their emotional instability. Every human being capable of empathy must be touched when seeing how much cannabis can help these children and their families. Now is the time to go the next step and add severe forms of autistic spectrum syndrome to medical cannabis qualifying conditions.[65]

[65] Thanks to Christian Bogner, M.D. for his wonderful summary of autism research and for sharing his personal experiences with his autistic son with me. You can find his manuscript "The Endocannabinoid System as it Relates to Autism" (with Joe Stone, 2014) here: http://de.scribd.com/doc/218971076/The-Endocannabinoid-System-as-it-Relates-to-Autism.

Marijuana, Reading, and Language Understanding

Many consumers of marijuana find that a high, especially a strong high, negatively interferes with their ability to read. Some report that they have problems concentrating on a text or they often seem to forget what they just read in a previous paragraph. However, in my opinion the negative effects described by these consumers depend on various factors: a strain that has been harvested too late contains more CBN (Cannabinol), a metabolite of THC that can lead to disorientation and short-term memory disruptions. Also, the route of administration matters; when cannabis is heated at high temperatures in a joint or a pipe the smoke usually contains more CBN and other chemicals that make you sleepy and confused than when heated at much lower temperatures in a vaporizer or baked in a cookie. Some strains of cannabis will be better to give a reader a clear 'heady' high which is helpful for concentration and thinking, whereas other strains affect the body more and can make a reader sleepy. Most consumers today use poor quality marijuana that has been produced under bad conditions, sold on a black market with virtually no quality controls. Often, the end consumer even obtains marijuana laced with lead or other substances that can actually be pretty harmful and certainly not best suited for the enhancement of cognitive abilities.

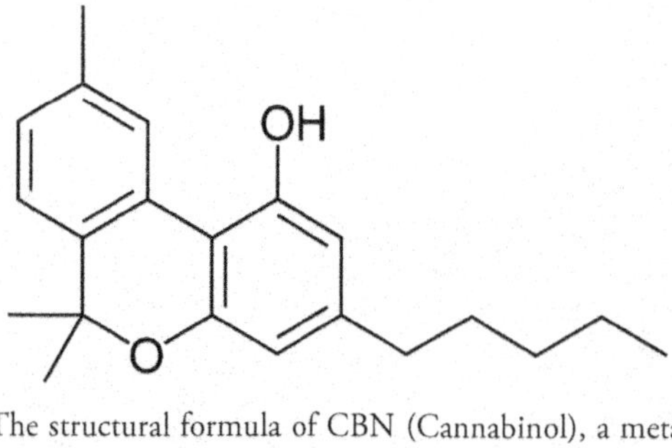

The structural formula of CBN (Cannabinol), a metabolite of THC that can make a marijuana consumer disoriented and confused.

The Facilitation of Reading During a High

Many other users have reported that marijuana actually helps them in the process of reading. Robert Burruss, a contributor to Lester Grinspoon's website project *marijuana-uses.com*, describes his former self as an effective illiterate at the age of 31. While at that time he was able to read single words and to pick up the gist from texts and to even get the meaning of some sentences, he writes that he never really quite understood the meaning of full sentences and texts written by others. One day, he sat down, puffed a joint and opened the book *Lady Chatterly's Lover* to "look for dirty words":

"I have no memory of the intervening moments before I learned to read. Perhaps only a few seconds passed. Maybe minutes. I don't know. All I recall is opening the book at a random place, or perhaps at multiple random places, and the next thing I know is that I'm walking up a stone path with flowers beside it, to the gardener's cottage, which has a thatched roof. The sky in the mental scene which the written words were creating is grayish, and the air is comfortably warm and slightly humid.
The sort of teleportation which the book and the joint provoked that night ... that was the first time in my life that mental images had been created by printed words. Until that night I had been unable to comprehend phrases longer than about three words. Until that night I had thought that everyone read that way, by looking at words and phrases and then fabricating an interpretation – highly personal, of course, though I didn't know it then – of the writer's intent. The seeing of mental images – and from printed words no less! – was the second great revelation of my life." [66]

We know from many other reports and studies that a marijuana high often enhances the process of imagination – be it visual, auditory, gustatory, olfactory or tactile imagination. Here, the ability to associate visual scenes to sentences seems to have helped the untrained reader Robert Burruss to finally come to a point where he could fully grasp the meaning of whole sentences. Enhanced imagery during reading has been reported by many other readers, who appreciate the marijuana high for giving them a more vivid imagin-

[66] Burruss, Robert (2010), "Lady Chatterly Stoned", in: Lester Grinspoon (ed.) (2013), marijuana-uses.com.

ing of the stories they read. Many others find that a high can also be helpful to focus their attention while they read. A clear "heady" high seems to help some users to focus their attention on a text without making them tired or loosing the thread of what they are reading.

The Enhancement of Foreign Language Understanding

Interestingly, this process of a "sudden switching" from tediously puzzling together the meaning of words to the sudden grasp of a sentence as a whole during a marijuana high has been reported by other users. "T.D." (anonymous author), a graduate student of Asian languages in his 30s with more than 10 years of experience as a marijuana user reports:

"My approach when studying stoned was always to bring as much concentration to bear on whatever aspect of the task I was working on, apply sustained effort until I had reached a conclusion, and then hurriedly write it down before forgetting it. Reaching the end of a sentence, I would then re-read all my notes and attempt to piece the meaning together. (...) On one particular occasion (...) something different happened. For some indeterminate period of time I was straining over a sentence and all at once, in a moment, the entire sentence as a single unit "flashed" in my mind and I read not syllable-by-syllable translated, but "read" the sentence as a coherent meaning unit.
Now it is usually hoped that at some point in the career of a foreign language specialist, this will happen. And, I am sure that there are those for whom this more "intuitive" approach to language comes naturally, and for whom strictly logical and rational thought seems painful and equally alien from everyday functional existence. But for me, going through life without a dominant framework of linear thought seemed to court danger, if not madness. Yet as a result of my experience, I could see in clearly demonstrable terms the facility of such occasionally less logically-stringent states-of-mind, in which the progression of thoughts is no longer logically sequential, but rather arise one after another through thematic association. It was in just such a state of mind where the marijuana which induced it had served as a catalyst to galvanize my comprehension of the language." [67]

---

[67] "T.D." (anonymous author), "Some Experiences with Language Facility and Learning", in: Lester Grinspoon (ed.) (2013), *marijuana-uses.com.*

Better Abilities To Understand Speech

In his study "On Being Stoned", Harvard psychologist Charles Tart found it to be very characteristic that marijuana users can understand the words of songs which are not clear to them when they are straight. Other marijuana users (including myself) have observed that during a high, they suddenly understand a spoken foreign language better (if they have some previous understanding of the language).

On the one hand, these enhancements might be the result of the effect of marijuana on our attention. It seems to be one of the most basic effects of marijuana to focus our attention, that is, to help with what cognitive scientists would call "selective attention" - focusing on a certain kind of pattern or object and screening out other perceptual stimuli. Thus, a high could facilitate our ability to focus on a song text and to screen out otherwise distracting musical stimuli, or to focus on the sound of spoken words of a different language and to screen out other noises and perceptual stimuli, such as visual distractions.

More importantly, however, an enhanced capacity for reading and understanding during a high might be subserved by a subtler enhancement of our pattern recognition abilities. I have discussed this before in my essay, "Marijuana, Pattern Recognition, and What It Means to Be High". Countless users of marijuana have observed that they suddenly see a new *gestalt* or, in other words, a pattern, during a high. Being high, a young man suddenly realizes that he is walking in a rigid way, or a woman sees a new pattern of insecurity in the behavior of her friend. In a similar way, the complete meaning, the "*gestalt*", of a sentence in the book *Lady Chatterly's Lover* suddenly pops up to the almost illiterate Robert Burruss while he is high.

It remains an open question whether the cannabinoids in marijuana could also directly affect cognitive processes underlying our language processing in a different, more specific way. So far, I have not seen any significant research concerning that issue, but I believe it might be a promising research focus for cognitive (neuro-) scientists.

Marijuana and Creativity. A Love Story

"The creative is the place where no one else has ever been. You have to leave the city of your comfort and go into the wilderness of your intuition. What you'll discover will be wonderful. What you'll discover is yourself."

Alan Alda, Actor, Author, and Filmmaker

In a good relationship, a loving partner can inspire you in many ways. She (or he) may simple inspire you on a daily basis with their beautiful smile, encourage you to compose your music or help you to relax and refresh your energy. Your beloved partner could also help you to discipline yourself, to keep focused on writing a book, drag you to a movie she finds interesting for you to help in collecting ideas for a movie project, or provide valuable feedback during the creation process of your new musical composition. There are many ways a lover can help your creative output. Naturally, things can also go wrong. Your partner may also negatively influence you or completely block your creative output in just as many ways. He or she could distract your attention away from your project, undermine your self-esteem or keep you constantly busy with irrelevant problems.

Marijuana can be to your creativity like a loving partner; a muse, inspiration and help in many ways. But like a partner, it can also interfere negatively with your creative output in many ways when things go wrong in the relationship.

Does marijuana "enhance creativity"? The short answer is that marijuana certainly has a magnificent potential to do so. However, that does not mean that marijuana automatically pushes a "creative boost button", for a simple reason; there is no such button. Creativity is not like electricity or a mysterious energy. It is the product of a successful coordination of various highly complex cognitive skills – and often crucially involves an additional coordination of many physical skills. This basic fact is often overlooked in discussions when it comes to marijuana and creativity – not only by laymen, but also by scientists. The marijuana high affects various cognitive processes that are part of the creative process. If we want to know how marijuana can enhance creativity, we have to look at a variety of cognitive alterations during a marijuana high and analyze how these can be either helpful or, otherwise, detrimental for a creative activity.

A Variety of Creative Processes, Phases, and Cannabis Strains

The story gets even more interesting, though, when we look at a variety of creative processes. Playing a new drum solo in a band during a live performance demands the coordination of various cognitive abilities, non-verbal communication, and, simultaneously, flawless hand-eye coordination. These abilities are very different from those a poet needs when he silently sits down trying to write a poem about the sound of wind going through a field of rye. For the spontaneous exploration of dance, a solo dancer requires the perfect functioning motor control of his whole body, timing and an ongoing stream of ideas in how to transpose music into motion.

The marijuana user and drummer genius Gene Krupa, circa 1946

Marijuana can indeed help artists, musicians and others by a multitude of cognitive alterations that enhance a whole variety of cognitive processes. But we shouldn't expect this to be a simple relationship. The many alterations of cognitive processes during a marijuana high have a different impact on various creative processes. Moreover, there are not only vastly different creative processes, but also, creative processes come in different phases. In his book *High. Marijuana in the Lives of Americans*, William Novak quotes an essayist stating:

"I just can't write well on grass. My grammar and syntax get screwed up, and I can get caught in the details. I do some of my thinking stoned, and the more linear work is done straight."[68]

---

[68] Novak, William (1980), *High Culture: Marijuana in the Lives of Americans.* Massachusetts: The Cannabis Institute of America, Inc, p.138.

If you want to use marijuana for creative purposes you will have to find out whether marijuana can help you for a certain creative process during a certain phase. As more advanced marijuana users know, this will also depend on the type and strain of cannabis you are using. Often, we forget another crucial aspect at this point; whether marijuana can help you for a certain creative process also crucially depends on your own psychological makeup. While a person with Obsessive Compulsive Disorder (OCD) may appreciate that cannabis helps him loosen up a bit and to break with routines, a user with a tendency for ADHS (Attention Deficit Hyperactivity Syndrome) might similarly appreciate the fact that certain strains can calm him down and help to focus his attention better. Others may find such relaxation effects as being counterproductive for their work. In an interview with HIGH TIMES Magazine, essayist Susan Sontag once said that for writing, marijuana would relax her too much – she said she preferred a little speed once in a while.[69] Other writers claim they can write perfectly during a high because they feel more concentrated and have a better flow while writing.

Essayist Susan Sontag, 1933-2004

In what follows, I will name a few effects of marijuana and outline how they might impact creativity; the list is not supposed to be complete. My intention is to broaden the perspective of many researchers in the field who have tried to reduce the "creativity-boosting" power of marijuana to one or two effects of marijuana that have been confirmed in psychological studies (as for instance "hyper-priming").

The High: A Variety of Cognitive and Perceptual Alterations

The taste of marijuana has often been underestimated because it is considered a "weed" and a "drug", and because it is often used in the wrong way mixed with bad tobacco. Use a well-grown, great tasting strain in a vaporizer

---

69 Interview with Susan Sontag in the *HIGH TIMES Magazine*, March 1978, #31.

and you will experience the obvious: marijuana has a complex herbal, nutty, fruity, earthy, delicate taste. Similarly, the effects of marijuana are usually underestimated as just bringing a euphoric "high", relaxation, and something like a better associative flow in thinking. The story, however, just like with taste, is much more complex – and much more interesting.

Based on the analysis of hundreds of reports of marijuana users and studies thereof I have described several cognitive and perceptual alterations during a high leading to various enhancements in my study *High. Insights on Marijuana.*[70] Ultimately, my study aims to explain how these alterations can lead to creative insights. During a high, users experience a hyperfocused attention. Sensations of all kinds coming into that focus are not only experienced as more intense, but are also perceived with more detail.

Furthermore, while high the capacity of users to imagine situations is enhanced and they can often vividly recall episodes of their past, sometimes reviving long forgotten memories in astonishing detail. Marijuana users often describe how their mind races quickly through memories, ideas or associations. They report the enhancement for pattern recognition of various kinds during a high; seeing new patterns in an artistic painting style, in the playing style of a jazz musician or in the behavior of friends. Many have used the high for better introspection and have found out important aspects about themselves. Many others have reported an enhanced empathic understanding of others, feeling that they can more easily put themselves "into their moccasins". Other users report spontaneous insights during a marijuana high.

Briefly then, the marijuana high brings a variety of mind-alterations which can be positively used – but the successful use of a high for a creative process crucially depends on the knowledge and the ability of a user to choose the right strain under the right circumstances for the a certain phase of a certain creative process.

So, what are the effects of marijuana on cognition which can turn out to

---

[70] Marincolo, Sebastián (2010) *High. Insights on Marijuana.* Dog Ear Publishing, Indianapolis, USA 2010.

be especially helpful for creative processes, and how can they enhance a users creative activity? In the following, I will describe some of the typical cognitive alterations during a high used by artists and other people.

Hyperfocusing, Redirection of Attention, and the Moment of Awe

One of the most fundamental effects of a marijuana high is the often reported hyperfocus of attention; whatever comes into the attentional focus of the user becomes not only more intense, but can often be explored in more detail. Time seems to slow down. The stunning experience of the here and now often leads to an exciting feeling of awe and wonder. Often, subjects report that their attention is redirected to aspects of objects or situations they would usually not attend to that much. Suddenly, they do not look at an object anymore under pragmatic aspects – they do not think about the usual ways to use it, but start to explore it's colors, surfaces or shades from a different perspective, like an artist or a scientist scrutinizing its nature. The Irish poet W. B. Yeats once wrote about his experience during a strong marijuana high in Paris:

The Irish poet and cannabis user William Butler Yeats (1865-1939)

"I opened my eyes and looked at some red ornament on the mantel-piece, and at once the room was full of harmonies of red, but when a blue china figure caught my eye the harmonies became blue upon the instant. I was puzzled, for the reds were all there, nothing had changed, but they were no longer important or harmonious; and why had the blues so unimportant but a moment ago become exciting and delightful? Thereupon it struck me that I was seeing like a painter, and that in the course of the evening every one there would change through every kind of artistic perception."[71]

For many marijuana users, this feeling of

[71] William Butler Yeats (1906), *Discoveries*, chapter „Concerning Saints and Artists" (quotation taken from veryimportantpotheads.com).

awe and wonder can be a crucial and life-changing catalyst for deeper investigations into patterns they have not explored before. In his essay "Creativity, Marijuana and a 'Butterfly Effect in Thought'", Jason Silva points out the importance of the feeling of wonder induced during a high:

"(...) Marijuana enhances our ability to marvel: In some mysterious and uncannily recurring way, marijuana can induce an almost 'synesthetic ecstasy,' whereby a loosening of the usually firm borders that separate our five senses allows for a broader, deeper, more profound, and often time-dilated "interpretation" and "internalization" of moment-to-moment experience."[72]

Synesthesia, Pattern Recognition and Metaphors

Jason Silva mentions the effect of *synesthesia.* Marijuana users experience full-blown synesthetic effects like seeing colors while listening to a guitar solo only for higher doses of marijuana. I have argued before that a pre-synesthetic effect of marijuana occurring even at lower doses may be constitutive for creative enhancements, leading to what I have called an enhanced ability for "*isomorphism* extraction".[73] In other words, the high seems to enhance our ability to extract similarities between seemingly unrelated concepts, objects, processes or patterns. The process of pattern recognition itself always involves the extraction of a similarity; when you look at a painting that you have never seen before and you recognize it as by Edgar Degas, you extract a similarity between the painting you are looking at and other Degas paintings works you have seen in the past; you *re-cognize* the style on the basis of an isomorphism extraction, a cognitive process to find similarities. Many marijuana users have reported that they come to a clear perception of the patterns of their own daily habits, which helps them to step outside the box and to break with these habits and routines. William James once said: *"Genius means little more than the faculty of perceiving in an unhabitual way."*

[72] http://realitysandwich.com/113157/marijuana_effect/, 2012

[73] Compare Marincolo, Sebastián (2010), *High. Insights on Marijuana.* Dogear Publishing, Indiana, USA. My argument is based on the groundbreaking work of the neuroscientists Vilayanur Ramachandran and Edward Hubbard.

Many reports from marijuana users show that during a high, they attend to different aspects of situations, objects or behaviors and discover new patterns. Suddenly they link seemingly unrelated structures, or concepts, a process that is crucial for the invention of new metaphors.[74] Obviously, this is one of the fundamental effects of marijuana used by creative people in many areas.

A typical Edgar Degas painting of dancers, circa 1888

Enhanced Imagination

There are countless reports of marijuana users regarding vivid imagery during a marijuana high and their enhanced ability for imagination. The effect usually goes undisputed in conversations about marijuana, but it is often overseen how important it is for various cognitive skills and especially for creative purposes. A New York painter reports:

"What I will allow myself to do – and succeed quite well in doing – is to paint mentally while I'm stoned. An image comes to my head, and, I refine it, rearrange it a number of times, and then let it float. The final stage of the image comes back to me, in a flash, later, when I'm straight. I can then use the mental painting as a series of shortcut steps."[75]

---

[74] The effect of semantic hyperpriming is presumably based to the pre-synesthetic effect described. Hyperpriming is a process in which subjects make unpredictable connections between items – a much-discussed recent study has shown that marijuana can induce a state of hyperpriming in subjects. See Morgan, CJA; Rothwell, E; Atkinson, H; Mason, O; Curran, HV; (2010), "Hyper-priming in cannabis users: A naturalistic study of the effects of cannabis on semantic memory function", *Psychiatry Res*, 176 (2-3) pp.213 - 218.

[75] William Novak (1980), *High Culture. Marijuana in the Lives of Americans*, The

Naturally, imagination is not restricted to the visual area; chefs have reported how they can better imagine the taste of combinations of food during a high, and musicians describe an enhanced ability to imagine melodies played by various instruments, thereby aiding their composing skills.

The Enhancement of Empathic Understanding

Ralph Waldo Emerson once said: "A painter told me that nobody could draw a tree without in some sort becoming a tree; or draw a child by studying the outlines of its form merely but by watching for a time his motions and plays, the painter enters into his nature and can then draw him at every attitude..."

Marijuana users often experience an enhancement of their empathic skills during a high. This enhancement is never named in connection with creativity, but I think it can be of fundamental importance when it comes to various creative processes. Writers may feel that they are better able to slip into another personality to, for instance, write fiction involving the feelings of their characters. Generally, an enhanced ability to see the world through the eyes of others can be of tremendous help when attempting to come up with new perspectives and ideas. Obviously, this enhancement can also help actors to better slip into their roles and to create a new persona.

Marijuana and Creativity: A New Approach

Let me conclude with some remarks about existing research on marijuana and creativity. Many studies on this subject will have to be critically re-evaluated. Scientists usually work with inexperienced test subjects who do not necessarily know how to use marijuana correctly for the purpose of creative enhancements. But as argued above, refined knowledge and experience with marijuana is crucial for a user to profit from various cognitive alterations during a high. Also, many studies focus on a simplified and abstract

---

Cannabis Institute of America, Massachusetts, p.131.

definition of creativity. I have argued here that the cognitive skills involved in creativity can vary significantly given the diversity of creative processes. Future research concerning the effects of a marijuana high on creativity should be directed at finding out about the basic cognitive effects of various cannabinoids and then investigate how these influence a diversity of creative activities under favorable circumstances – ranging from an improvised trumpet solo to the writing of a poem.

Personal Transformation with Marijuana

"Muta {slang term for marijuana} takes all the goddamn hardness and evil out of you, cuts down the tush-hog bullying side of your personality and makes you think straight, with your head instead of your fist; it digs the truth out and dangles it right in front of your nose. Everything comes out in the wash starched and clean. A viper doesn't like lies –he's on the up-and-up and makes you get on the ground floor with him."
Milton "Mezz" Mezzrow, Really the Blues, 1946

In recent years, medical marijuana has been increasingly used as treatment for Post-Traumatic Stress Disorder (PTSD; in some states in the U.S. such as California, approximately 25% medical marijuana patients have a prescription for PTSD. Israel's government also treats traumatized soldiers with cannabis. We know from numerous reports from war veterans and other PTSD sufferers that marijuana can give instant relief and can actually help to minimize symptoms of PTSD in the long term. We still don't know exactly how marijuana helps these patients, but we have some clues. About a decade ago, physiologist Beat Lutz found that the endocannabinoids in our brain play an important role in the extinction of aversive memories.[76] Inhaled or ingested (exogenous) cannabis may act on this system and help PTSD patients to overcome their traumas and to move on and personally develop. A recent study with rats by the Israeli scientists Eti Ganon-Elazar and Irit Akirav[77] suggests that cannabinoids do not really extinguish aversive

Photo from World War I. The Australian soldier left shows post traumatic symptoms from a shell-shock, the often called "thousand yards-stare."

[76] Lutz, Beat (2002), „The Endogenous Cannabinoid System Controls Extinction Of Aversive Memories", *Nature, August 1, 418(6897): pp.530-4.*

[77] Ganon-Elazar, Eti, and Akirav, Irit (2011) "Cannabinoids Prevent the Development of Behavioral and Endocrine Alterations in a Rat Model of Intense Stress", *Neuropsychopharmacology*;|doi:10.1038/npp.2011.204

memories – the memories still seem to be there, but they do not cause such extreme reactions. So, marijuana probably helps patients to overcome their traumas without extinguishing their episodic memories of the trauma they suffered.

Cognitive Effects of a Marijuana High

These findings on the neurological level are promising for the treatment of millions of traumatized people. In this essay, however, I want to explain further ways in which marijuana can be personally used or used in psychotherapy. I will argue here that a marijuana high and its cognitive effects can help many more ways of positive personal transformations – and we have a lot of anecdotal evidence for this claim. For example, here is one report from a marijuana user who used marijuana to get away from a shallow lifestyle in a marriage that was almost wrecked:

> *(…) we've discovered that what we both really want is contentment with what we have rather than the perfectionism of accumulation and put-ons. Marijuana has landed us solidly into a groove of change; it's broken down barriers between us, and probably saved our marriage and family. We have a heightened sense of what we're all about as a couple, we're better parents (…).*[78]

Mindsight and Marijuana

How could the marijuana high help with such a positive and desired personal transformation? Some of the answers may lie in a relatively new approach in psychotherapy suggested by Daniel Siegel, clinical professor of psychiatry at the UCLA Medical School. In his book *Mindsight. The New Science of Personal Transformation,*[79] he combines Western psychotherapy and new findings from the neurosciences with Eastern meditation practices.

---

[78] "E. Cleaves" (2010) "We're Not Bluffing Anymore", in: Grinspoon, Lester (ed.) (2013), *marijuana-uses.com.*

[79] Siegel, Daniel J. (2010), *Mindsight. The New Science of Personal Transformation,* Bantam; Reprint edition.

("Mindsight" is a term coined by Siegel to name our ability to understand our own and other minds.)

One of Siegel's core concepts is *integration,* which for him serves as the basis for happiness. He identifies eight areas of integration, but for the sake of brevity, I will mention only three here: *horizontal integration*, *vertical integration*, and the *integration of episodic memories*. Let me describe these briefly. Stuart, aged 93, was a patient of Siegel and came to him because he realized that he felt nothing when his wife got severely sick. It turned out that Stuart had never had such feelings because his parents had been emotionally cold with him. As a child, Stuart had probably "switched off" the right brain hemisphere, which is responsible for holistic thinking, non-verbal communication, and empathy. Instead, he used mainly his rational, logical and verbal left hemisphere and worked as a lawyer. Siegel successfully used various methods to direct Stuart's awareness to the processes in his right hemisphere. He helped Stuart to heal by integrating his right hemisphere – that's the process Siegel calls *horizontal integration.*

Siegel's patient Anne had "switched off" her feelings for her body as an 11-year-old child because she had decided to never feel again after the severe shock when her mother died. Here, Siegel realized that she needed to *vertically integrate* the feelings for her body again and he used various physical trainings and awareness exercises to help her to do so – with much success.

Traumatized patients often completely forget the traumatic episodes they experienced, such as rape, abuse, physical violence or a fatal loss. One of the most important techniques for Siegel's method of "mindsight" is to let his patient meditate and to let them find a secure space in the here and now. If they feel stressed, he asks them to imagine being at their favorite place, maybe a beach or the house of a favorite aunt during childhood. In this state, he can actually lead them to remember less fearfully a previously suppressed traumatic experience; which they need to better deal with their memories later. This is what Siegel calls the *integration of episodic memories.*

Siegel cites impressive studies to show that becoming aware of emotional traumas and training one's awareness over a longer period of time can lead to a real physical transformation in the brain because of neuroplasticity - the technical term for the fact that throughout our lives new synaptic connections and neurons can be grown. Our brain, therefore, is in a constant process of transformation.

Marijuana, Transformation, and Psychotherapy

I have argued in many places that a marijuana high can be temporarily used as a cognitive enhancer in numerous ways. I will now list some of them and comment how these could help patients – but also other marijuana users – to heal or to literally transform themselves into more healthy personalities.

One of the most fundamental effects of marijuana seems to be the hyperfocusing of attention. It is obvious that this effect can help users to focus on the here and now. Indian sadhus have used cannabis derivates for hundreds of years for meditative purposes. Marijuana can certainly help beginners to go deeper in their meditation or to focus on episodic memories.

Two sadhus near Pashupatinath temple in Kathmandu, Nepal.
Photo © Luca Galuzzi 2006, Wikimedia Commons

The enhancement of episodic memory retrieval during a high is one of the main effects described in detail by many marijuana users and could also help them to become aware of and integrate episodic memories. Furthermore, marijuana helps users to vividly imagine various situations, which could be generally helpful during a guided meditation.

In brief, then, the high can be helpful for the process of meditation – which can generally be used with the 'mindsight' method. The enhancement of episodic memory during a high can help in the process of the integration of episodic memories. Also, many users reported various enhancements of their *interoception* – the perception of one's own body. This could also help patients to 'get in touch' with their bodies, which facilitates the process of vertical integration. Finally, we have many reports from users indicating that marijuana seems to enhance several functions which are dominantly performed by the right hemisphere of the brain – imagery, empathic understanding, insights – so it could also help with horizontal integration.

Lester Grinspoon's amazing collection of essays and reports from marijuana users (not abusers)[80] shows that many have successfully used marijuana for introspection, insights, a better empathic understanding and various other cognitive processes, which often changed and positively transformed their lives. Siegel's approach shows us that psychotherapists should at least consider marijuana as an important tool to help millions of patients to overcome serious traumas, deep fears or depression based on previous processes of disintegration - and as a general means to facilitate a positive personal transformation.

---

[80] Grinspoon, Lester (ed.) (2014), *marijuana-uses.com.*

Part III Inspire

Vipers, Muggles, and The Evolution of Jazz

"I'm the king of everything
Got to get high before I sing
Sky is high, everybody's high
If you're a viper..."
'Viper's Drag' (1934), by Fats Waller

Without doubt the history of jazz and the use of marijuana are intimately intertwined. The story does not begin in the early 20th century in New Orleans, the birthplace of jazz. It begins hundreds of centuries earlier. Jazz and blues are deeply rooted in the African musical tradition, and the use of marijuana for medical, inspirational, religious and other purposes had been widespread and deeply influential in African societies and their cultural lives for decades. Presumably, the psychoactive effects of marijuana already played an important role in the way many African musical traditions had evolved. In his magnificently researched book "Smoke Signals", author Martin A. Lee writes:

"Black Africans employed a wide variety of devices (...) for inhaling 'dagga', as marijuana was called by several tribes, who regarded it as a "plant of insight. (...)Thrown upon bonfires, marijuana leaves and flowers augmented nocturnal healing rituals with drum circles, dancing, and singing that involved the spirit of the ancestors and thanked them for imparting knowledge of this botanical wonder."[81]

Cannabis and its psychoactive use made its way to South-, Middle- and North America through the slave trade of more than 11 million African slaves. Lee reports about how African slaves gathered at Congo Square in early nineteen-century New Orleans to sing and dance their music, a rite that would later become prohibited because slave owners were afraid that their slaves used their complex rhythms to communicate rebellious messages

---

[81] Lee, Martin A. (2013) Smoke Signals, *A Social History of Marijuana – Medical, Recreational, and Scientific*. Scribner, New York, p.14.

to each other.[82] However, the African traditions survived and would later influence the evolution of jazz.

In addition to some existing cannabis use in the U.S. amongst black slaves, thousands of Hindu immigrants from India brought the use of cannabis to the West Indies in the 1870s. From there, black and Mexican sailors picked up the habit and introduced marijuana use to the harbor of Storyville, the red light district of New Orleans, the city usually considered to be the birthplace of jazz. At the beginning of the 20th century, countless black jazz musicians performing in the bordellos of Storyville and other locations in New Orleans smoke what they call 'gage', 'tea', 'muggles', 'muta', 'Mary Jane'. They will soon call themselves 'Vipers' – allegedly named after the hissing sound taking a quick draw at a joint, or, as they began to call it, a 'reefer'.

Postcard of Storyville, New Orleans, around 1908

Working long night shifts, many vipers prefer smoking marijuana to alcohol. It does not give them the hangovers associated with excessive alcohol consumption. Louis Armstrong, born in 1901, grew up in extreme poverty of Storyville in a rough neighborhood known as "The Battlefield". A proud *Viper* himself from his youth and for all of his life, the first black international superstar will later reflect that a sequel to his biography might well be about "nothing but gage". Armstrong will remember about his use of marijuana:

"First place it's a thousand times better than whiskey … It's an Assistant – a friend a nice cheap drunk if you want to call it that …Good (very good) for Asthma – relaxes your nerves …"[83]

---

[82] *Ibid*, p.9.

[83] Armstrong, Louis (1999), *In His Own Words, Selected Writings*, Oxford University Press, p.114.

"We always looked at pot as a sort of medicine, a cheap drunk and with much better thoughts than one that's full of liquor."[84]

Along with jazz, the use of marijuana spreads to bigger cities like Chicago, Detroit, and New York. Around 1930 during the prohibition of alcohol – while marijuana is still legal – there are countless illegal 'speakeasies' serving alcohol, but also around 500 tolerated 'tea pads' (marijuana bars) in New York alone offering joints for around 20 cents. In the 1930s, Viper songs celebrating the use of marijuana become the rage of the jazz world, including 'Muggles' (Louis Armstrong), "Sweet Marijuana Brown" (Benny Goodman), Viper Mad (Sydney Bechet), "That Funny Reefer Man" (Cab Calloway), "Viper's Drag" (Fats Waller), or "Gimme a Pigfoot" (Bessie Smith).

The open reefer party, though, will soon be over for the Vipers. From the early days in New Orleans, white officials were not at all enthusiastic about this new self-confident, vibrant black jazz culture finding its way into the heart of white audiences. The first prohibitive laws against marijuana in the southern states are clearly targeted against its users - Hispanic immigrants and black musicians. Harry G. Anslinger, the nation's drug czar, openly uses outrageous racist claims to run his campaign against marijuana and to successfully justify a nationwide prohibition in 1937. In a Senate hearing on marijuana in 1937 he infamously states:

"There are 100,000 total marijuana smokers in the U.S., and most are Negroes, Hispanics, Filipinos and entertainers. Their Satanic music, jazz and swing, result from marijuana use. This marijuana causes white women to seek sexual relations with Negroes, entertainers and any others. ... The primary reason to outlaw marijuana is its effect on the degenerate races. Marijuana is an addictive drug which produces in its users insanity, criminality and death."[85]

---

[84] Jones, Max, and Little, John Clifton (1988) *Louis. The Louis Armstrong Story 1900-1971* DaCapo Press.

[85] Statement of Anslinger, H.J., Commissioner of Narcotics, Bureau of Narcotics, Department of the Treasury, http://www.druglibrary.org/schaffer/hemp/taxact/anslng1.htm.

The Multi-faceted Influence of Marijuana on Jazz

How much did marijuana really influence the early evolution of jazz? Most of Anslinger's claims above are of course ridiculous, but arguably, he was not all that wrong in seeing a connection between marijuana use and jazz. Many historians clearly see this connection, but most usually play down the profound and multi-faceted influence of the marijuana high on the early evolution of jazz. First of all, they underrate the complexity of the marijuana high and its many diverse effects on the performance of musicians. Second, another aspect often completely neglected by historians is that the marijuana high affected not only individual performances, but was also crucially involved in the evolution of the sub-cultural new lifestyle of which jazz was an expression. Let us look at the latter claim first.

Marijuana and the Viper Culture

Life is extremely hard for black citizens in the 1920s and 1930s in the U.S. The Jim Crow racial segregation laws in the South are still in force and will be up until 1965. They do not only mandate a segregation of black and white citizens in public schools and other public places, but have countless other repressive regulations restricting the liberties of black citizens. David Pilgrim reminds us about the thinking behind the Jim Crow 'system':

Ku Klux Klan members in Washington D.C. in 1928

"(...) The Jim Crow system was undergirded by the following beliefs or rationalizations: whites were superior to blacks in all important ways, including but not limited to intelligence, morality and civilized behavior; sexual relations between blacks and whites would produce a mongrel race which would destroy America; treating blacks as equals would encourage interracial sexual unions; any activity which suggested social equality encouraged interracial sexual rela-

tions; if necessary, violence must be used to keep blacks at the bottom of the racial hierarchy. "[86]

Black musicians are constantly humiliated by racial segregation and repression, and many of them go through extremely traumatizing experiences. Violations of the Jim Crow laws or norms – like stepping on the shadow of a white man – are punished with incredible violence; thousands of black citizens are lynched between 1882 and 1968, most of them in the South. Many emigrate to cities in the North, but starting a life there isn't quite a stroll in the park either, especially not for black musicians moving to the bigger cities to establish a career. In his autobiography *Really the Blues*, jazz clarinetist Milton "Mezz" Mezzrow recalls:

Milton "Mezz" Mezzrow, 1899-1972

"(...) it often happened that a man who migrated into town couldn't eat unless his woman made money off of other men. But these people didn't get nasty about it; many a guy kept on loving his woman and camping outside her door until she could let him in (...)"[87]

The musical tradition of the blues had always helped the black community come to terms with their hostile life conditions; it expressed the sadness of many, but also created a space in which they could regain strength, faith, and joy. As Milton Mezzrow put it:

"These blues from the South taught me one thing: You take off the weight off a good man a little and his song will start jumping with joy. "[88]

---

[86] Pilgrim, David (2000/2012), „What was Jim Crow", www.ferris.edu/jimcrow/what.htm.

[87] Mezzrow, Mezz (1946/1990), *Really the Blues*, Souvenir Press, London, p.46.

[88] *Ibid.*

Marijuana for Post-Traumatic Stress Syndrome, Anxiety-Relief, and as a Euphoriant

For many traumatized and oppressed black musicians, marijuana takes some more weight off. We know today that medical marijuana is used very effectively to treat post-traumatic stress syndrome (PTSD). In the recent years, myriads of patients with PTSD in the U.S. have received medical marijuana; most of them report that it helps them better than any other medicine. A new study has shown that there are endocannabinoid receptors in the part of the brain that regulates anxiety and the fight-or-flight response; this could also explain why so many patients find that consumed marijuana can help them to reduce anxiety.[89]

Marijuana presumably helped many jazz musicians to deal better with their traumas, fears and culturally imposed restrictions. Accordingly, we can see the heavy use of many jazz musicians at least to some degree as self-medication. At age 11, Billie Holiday's neighbor attempts to rape her; at age 14, she reportedly has to work as a child prostitute in Harlem for $5 a client. Holiday starts smoking marijuana (and drinking alcohol) habitually before she is a teenager. Her big idol, and life-long marijuana user, Louis Armstrong, has to work as a young boy to support his mother, but can not prevent her from also having to work as a prostitute. Holiday's other idol Bessie Smith, the Queen of Blues – who, like

Billie Holiday in the Downbeat Club 1947

89 Snyder, William (March 6, 2014), "Discovery sheds new light on marijuana's anxiety relief effects."http://news.vanderbilt.edu/2014/03/discovery-sheds-new-light-on-marijuana-anxiety-relief-effects/

Armstrong and Holiday, also used marijuana throughout her career – grows up in extreme poverty, is an orphan aged nine and starts to work on street corners to escape poverty.

Consumed cannabis can influences the endocannabinoid system in mood regulation, and there is another additional cognitive effect of the high which helps users with stress relief. Also, while high, users strongly focus their attention, often dwelling in the here and now, forgetting about past troubles or future problems. To say it with Cab Calloway's 1932 viper song "The Man from Harlem": *"I've got just what you need. Come on, sisters, light up on these weeds and get high and forget about everything."*

Louis Armstrong, Aquarium, New York 1946

Furthermore, we know that some marijuana strains can lead to euphoria during a high. Louis Armstrong remembers in his biography:

"It makes you feel good, man. It relaxes you, makes you forget all the bad things that happen to a Negro. It makes you feel wanted, and when you are with another tea smoker it makes you feel a special sense of kinship."

The Empathic Effect of Marijuana

The 'special kinship' mentioned by Armstrong adds another important aspect to the picture. When thinking about the hippie era, we usually consider it a fact that a marijuana high made users more loving and empathic. We tend to forget that it had a similar effect for many musicians and their audiences in the swing era of the 'Roaring Twenties – which also helped to pave the way for the later 1950s Beat Generation. Louis Armstrong goes on to explain the nature of the kinship between the Vipers:

"One reason we appreciated pot, as y'all calls it now, was the warmth it always brought forth from the other person – especially the ones that lit up a good stick of that shuzzit or gage (...)."[90]

In a similar vein, Mezz Mezzrow explains about jazz musicians using marijuana:

"We were on another plane in another sphere compared to the musicians who were bottle babies, (...) we liked things to be easy and relaxed, mellow and mild (...) their tones became hard and evil, not natural, soft and soulful (...)"[91]

The empathic effects of marijuana probably also help the democratization of music which plays a crucial part in the early evolution of jazz. Strong empathy is an equalizer: hierarchies become less important; solos are not only restricted anymore to singers or the classic solo instruments such as guitar or saxophone. As jazz pianist Herbie Hancock will later put it in his statement on jazz: *"It's not exclusive, but inclusive, which is the whole spirit of jazz."* Racial boundaries and prejudices are more easily overcome. Mezz Mezzrow, the white Viper from a Jewish family who famously sells the good quality "mezzroles" (joints) to other musicians, declares himself to be black – out of sympathy for black lifestyle and music.

The Effects of the Marijuana High on Musical Performance

So, our current knowledge concerning the medical uses of marijuana and reports from Viper jazz musicians strongly suggests that their use of marijuana played a positive role in the evolution of an early jazz culture. But can a high also positively influence the performance of a jazz musician? And if so, how? The effect most often cited as an answer, is the altered sense of time during a high. Dr. James Munch, pharmacologist and associate of Harry G.

---

[90] Jones, Max, and Little, John Clifton (1988) *Louis. The Louis Armstrong Story 1900-1971,* DaCapo Press.

[91] Mezzrow, Mezz (1946/1990), *Really the Blues,* Souvenir Press, London, p.94.

Anslinger during the 1930s and '40s, produced many ridiculous claims about the alleged horrible effects of marijuana, but clearly expressed this point years later, when he said about musicians using marijuana:

"(...) the chief effect as far as they were concerned was that it lengthens the sense of time, and therefore they could get more grace beats into their music than they could if they simply followed the written copy. (...) if you are using marijuana, you are going to work in about twice as much music between the first note and the second note. That's what made jazz musicians."[92]

Hyperfocusing, Mind Racing, and an Altered Sense of Time

Munch's point about the altered sense of time and its role in jazz music is important; however, this is only one of several crucial effects of marijuana which can play a positive role for jazz performers. If we want to understand them better, we have to look at the way these effects are interrelated. One of the fundamental effects of marijuana is to hyperfocus attention. Mezzrow remembers this hyperfocus for his auditory experience when he first got high:

"The first thing I noticed was that I began to hear my saxophone as though it was inside my head, but I couldn't hear much of the band in back of me, although I knew they were there. All the other instruments sounded like they were far off in distance;(...)". [93]

This hyperfocusing allows him to concentrate more fully on his immediate tactile sensation of his instrument, which improves his control:

"Then I began to feel the vibrations of the reed much more pronounced against my lip (...) I found I was slurring much better and putting just the right feeling into the phrase." [94]

---

92 Sloman, Larry "Ratso" (1998), *Reefer Madness. The History of Marijuana in America,* St. Martin's Griffin, New York, pp.146-147.

93 Mezzrow, Mezz (1946/1990), *Really the Blues,* Souvenir Press, London, p.72.

94 *Ibid.*

During a high, the hyperfocusing of attention not only allows for a more analytic perception of whatever it is directed at, the narrowing down of informational processes in the attentional focus probably also leads to mind racing, which may then lead to an altered sense of time.[95] In his 1938 report *Marihuana, America's New Drug Problem*, R. P. Walton states:

"The exaggeration of the sense of time is one of the most conspicuous effects. It is probably related to the rapid succession of ideas and impressions which cross the field of consciousness."[96]

The connection between the prolonged sense of time and 'mind racing' seems intuitively plausible and is backed by the experiential reports of many marijuana users throughout history. In his 1877 report about a marijuana high, the French doctor Charles Richet explained:

Cannabis Indica Leafs, Photo © Sebastián Marincolo 2012

"Time appears of an immeasurable length. Between two ideas clearly conceived, there are an infinity of others undetermined and incomplete, of which we have a vague consciousness, but which fill you with wonder at their number and their extent. With hashish the notion of time is completely overthrown. The moments are years, and the minutes are centuries; but I feel the insufficiency of language to express this illusion, and I believe, that one can only understand it by feeling it for himself."[97]

---

[95] Compare Marincolo, Sebastián (2010) *High. Insights on Marijuana*, chapter 6, "Intensified Imagination, Mind Racing, and Time Perception Distortions", Dog Ear Publishing, Indiana.

[96] Walton, R.P. (1938), *Marijuana. America's New Drug Problem*, Philadelphia, Lippincott, p.105.

[97] Quoted in Malmo-Levine, David (2003), *Cannabis Culture*, Dec. 5., http://www.cannabisculture.com/articles/3075.html.

Richet mentions an infinity of ideas between two ideas – we can take his "ideas" to refer to what we would usually call thoughts today. The acceleration of a stream of "ideas" or thought processes is sometimes experienced as a stream of associatively connected thoughts, memories or imaginations, much depending on the dosage consumed. Obviously, the acceleration of mental processes in a narrowed down tunnel of attention can help a musician to more rapidly play an improvised solo, or to keep up with the speed of fellow musicians. While this acceleration during a high probably leads to a musician's ability to "*work in about twice as much music between the first note and the second note*", it can also hinder musicians in keeping time with others. We will come back to this particular downside later on. Beforehand, we should look at some other effects of the marijuana high which may have had a positive effect on the musical performance of the Vipers.

Short-Term Memory Disruptions, Enhanced Pattern Recognition, and Imagination

With their minds unusually focused to the present moment or thought, marijuana users sometimes forget about the original subject framing the discourse. This often leads to a "what were we talking about?" moment when we are losing the thread of the conversation. Whereas inexperienced users – especially when using high dosages of certain strains – become disoriented, skilled users who consume certain other strains of good quality marijuana can keep the thread. Still, their stream of thought is usually less constrained by the original theme or frame where it started, and also less constrained by the goal where it was intending to go. The stronger concentration on the 'here and now' allows a stream of thoughts or imaginations to 'jump' more freely along unusually wider associations.

Many marijuana users have also reported an enhanced ability to see new patterns during a high and they often find new similarities between various patterns. In a musical performance, these effects can then lead to a rapid improvisation over known musical themes, which loosely associate them to new patterns and ideas or they can also lead to new connections or transitions between musical themes. Subjectively, this leads to the feeling of a rapid and effortless flow of ideas.

Furthermore, innumerable marijuana users have described that a high enhances their imaginative abilities – visually, auditory, gustatory or otherwise – a capacity which is crucial for the production of new ideas. It goes without saying how important an enhanced ability for auditory imagination could be for a musical performer coming up with a new improvisation on stage, or for a composer working on a new song.

In Milton Mezzrow's first high, the interrelated effects of a hyperfocus of attention, mind racing, an enhanced pattern recognition ability and an enhanced imagination lead to a smooth, imaginative flow in his playing:

"*All the notes came easing out of my horn like they'd already been made up, greased and stuffed into the bell, so all I had to do was to blow a little and send them on their way, one right after the other, never missing, never behind time, all without an ounce of effort. The phrases seem to have continuity to them and I was sticking to the theme without ever going tangent. I felt I could go on playing for years without running out of ideas and energy*".

Marijuana as an Aphrodisiac

Apart from these interrelated cognitive effects of the marijuana high, Mezzrow mentions another interesting and important way in which marijuana affected jazz:

"*Us vipers began to know that we had a gang of things in common: (...) we all decided that the muta had some aphrodisiac qualities too, which didn't run us away from it.*"[98]

Many commentators who cited Mezzrow's passages about the influence of marijuana on his performance on stage usually forget to mention that his

---

[98] Mezzrow, Mezz (1946/1990), *Really the Blues, Souvenir Press, London*, p.93.

high adventure on stage ends in an ecstatic group experience similar to scenes witnessed decades later at the height of Beatlemania:

Dancers at the Elk's Club in Washington, D.C., 1943

"The people were going crazy over the subtle changes in our playing; (...) some kind of electricity was crackling in the air and it made them all glow and jump. (…) it seemed like all the people on the dance floor were melted down into one solid, mesmerized mass; (…) looking up at the band with hypnotic eyes and swaying (…). An entertainer (...) was throwing herself around like a snake with the hives. The rhythm really had this queen; (…) what she was doing with (...) her anatomy isn't discussed in mixed company. "Don't do that!" she yelled. "Don't do that to me!"[99]

That is probably what Duke Ellington meant when he said about jazz: "*By and large, jazz has always been like the kind of a man you wouldn't want your daughter to associate with.*"

The Enhancement of Empathic Understanding

I have argued in various places that for skilled users, the marijuana high can lead to various cognitive enhancements, which can also lead to a fundamental enhancement of empathic understanding of others.[100] This effect goes

---

[99] *Ibid.*, p.73.

[100] For a detailed argument on how a marijuana high might positively affect empathic understanding compare (2010) *High. Insights on Marijuana*, Dog Ear Publishing, Indiana, and Sebastián Marincolo (2013), *High. Das positive Potential von Marijuana,*

way beyond stronger emotional bonds or the warm sense of sympathy between the Viper musicians mentioned above. The enhancement of empathic understanding during a high allows marijuana users to actually better understand what others are feeling and thinking.

Musicians with an enhanced empathic understanding of each other communicate better – both on and off stage. When performing live together, jazz musicians improvise and do not follow strict pre-meditated rules; their performance as a group is crucially dependent on their mutual understanding, reacting to each other within the flow of their performance. Empathic understanding is not only helpful for jazz musicians; it is absolutely crucial to their new and liberated form of music, which depends upon a mutual interplay between musicians spontaneously reacting to each other during their performances.

In the swing era of the 1930s, legendary Billie Holiday and Lester Young were known for their almost telepathic performances; both were experienced Vipers and used marijuana regularly. During the time of performing in the Cafe Society, Billie Holiday used to go on taxi rides between the sets to smoke some marijuana, because smoking marijuana wasn't allowed in the club.[101] Like many other jazz Vipers, their use of marijuana may have helped them to come to a closer mutual understanding. We will never know for sure in individual cases how much of a positive influence the marijuana high had on the mutual understanding of musicians; but from what we know about the influence of the high on empathic understanding and from personal reports of jazz musicians and their friends, it certainly seems like many jazz musicians profited from this effect of the high.

A Note of Caution

In an interview, jazz clarinetist and bandleader Artie Shaw said he once got frustrated with Viper Chuck Peterson, the first trumpet player in his band. Shaw felt that Peterson made the band lag when playing high. He confronted Peterson, who thought he was playing just fine and they came to

---

Klett-Cotta/Tropen, Stuttgart.

[101] See Clarke, Donald (2002), *Billie Holiday: Wishing on the Moon, DaCapo Press, p.160.*

a deal. Shaw, who had smoked marijuana for a while as a young adult suggested they perform high together – if that worked he said, they would turn on together every night from then on. Shaw reports:

"He gave it to me, I smoked it, and I was playing over my head. I was hearing shit I'd never heard before in those same old arrangements. I finished and turned to him. 'You win,' I said. 'No, man,' he said. 'I lose.'
He had been giving me incredulous looks during the evening and I thought he was thinking, 'Man, this guy is blowing his head off.' I was hearing great things. But the technical ability to do it – it's like driving drunk. You feel great, but you don't know what you're doing. At least he was honest about it.'[102]

Now, does this show that jazz musicians smoking marijuana were generally undergoing a delusion about their own performance during a high? Hardly. From all we know, experienced Vipers like Louis Armstrong, Bessie Smith, Billie Holiday, Lester Young, Cab Calloway, Fats Waller, Theloneous Monk, Anita O'Day, Lionel Hampton, Count Basie, Duke Ellington and many others were doing more than just fine performing under the influence of marijuana. Dizzy Gillespie, who wrote that he was turned on to pot when he came to New York in 1937, remembers in his autobiography that almost all jazz musicians he knew were smoking pot and some of the older musicians had been smoking pot for 40 or 50 years. Surely, they were not all victims of a self-delusion when it came to performing high.

Bessie Smith, (1894-37). Photograph by Carl Van Vechten 1936

[102] Saroyan, Aram (August 6, 2000), „Artie Shaw Talking“, *Los Angeles Times*, http://articles.latimes.com/2000/aug/06/magazine/tm-65218/2

However, Artie Shaw's story reminds us that not every musician can perform well while high; a true viper has to master the effects of the high and has to learn how to ride a high – just like a surfer has to learn how to ride a wave with a surfboard. Note, however, that even musicians who cannot deal with a marijuana high on stage, the high can still turn out to be helpful to them in other ways. Artie Shaw notes above that under the influence of marijuana, he was hearing things in old arrangements that he had never heard before when playing straight. This new perception could have also paved the way for him for new interpretations or arrangements. A marijuana high can help creative processes or activities in many ways and in various phases.

Likewise, a writer may feel that he can generate great ideas during a high, while actually feeling that the high doesn't really help or even strongly interferes with the process of actually writing down the details. If used in the wrong way, marijuana can certainly also have a negative influence on creative performances. The lesson to learn here is that generally, if you want to use marijuana positively for creative purposes, you have to learn how much of which strain can help you in a specific phase of a certain creative processes. And, as the Jamaican's say, some "don't have a head for ganja"; we all have to find out whether marijuana helps us for any kind of creative or other purpose.

The marijuana high enhanced Louis Armstrong's performance; he was an expert in riding a marijuana high, and he loved it. But that certainly does not mean that his musical ability can be reduced to the influence of marijuana. It was made possible by his enormous talent, his character, his discipline, training and experience. Likewise, the evolution of jazz certainly was not driven solely by marijuana use, but rather was made possible by many factors including the cultural mixture of African, European and Caribbean music and lifestyles, the sociological process of urbanization, amongst many other factors. The aforementioned red light area of New Orleans Storyville, for instance, played a big role in the early development of jazz: *"it was a rough area where white values of taste were absent. This made it easier for musicians to develop expressive techniques, slow tempos (for sexy, slow dances) and timbre variation."* [103]

---

[103] Devaux, Scott, and Giddins, Gary (2009) "Jazz", College edition online, chapter 4, 8. Storyville, http://www.wwnorton.com/college/music/jazz/ch/04/outline.aspx

But if we look at the whole bouquet of the now better-known cognitive changes during a high we come to understand that marijuana substantially contributed to the evolution of jazz. It helped countless skilled artists to repeatedly come up with new inventive solos, fluid, rapid and imaginative playing; it helped them not just in their individual performances, but also, to better understand each other and to "click" together on stage.

Off stage, the use of marijuana changed the thinking and lifestyle of many jazz musicians from an early age. *"If you don't live it, it won't come out of your horn"*, Charlie Parker once said. From the very beginnings in New Orleans, the marijuana high was integral to the early evolution of a free, joyous, empathic, rebellious, uninhibited, imaginative and creative viper subculture and lifestyle. They celebrated this lifestyle with their jazz – a radically new form of music, one of the greatest cultural achievements to come out of the U.S. and a lasting inspiration to people all around the world.[104]

[104] For a great source on cannabis and music take a look at Jörg Fachner's publications on the subject. Also, see Cronin, Russell (2004), „The History of Music and Marijuana", *Cannabis Culture Magazine.* www.cannabisculture.com/content/2004/09/08/History-Music-and-Marijuana-Part-One.

What Hashish Did To Walter Benjamin

It's the year 1927. In the U.S., marijuana use is widespread almost exclusively amongst Mexicans and black Jazz musicians during the swinging "Roaring Twenties". In Berlin, the decade is also in full swing, but the first signs of the upcoming catastrophic end of this era are already visible, as radical national socialist groups clash more and more with Marxists in violent street fights. Here, the popular drugs – outside from alcohol and tobacco – are cocaine and morphine, available on prescription only, but happily peddled by many doctors who are making good money pushing their illicit wares. The authorities aren't really enforcing the drug regulations; they protect the interests of the big drug-producing pharmaceutical companies in Germany, which are now part of the leading pharmaceutical industry in the world.

Berlin in the late 1920s.

Writers like Ernst Jünger and Gottfried Benn consume cocaine. Legendary naked dancer Anita Berber, who openly admits to consuming large amounts of cocaine, is the style-icon of the day. She steps out of a car on the Kurfürstendamm with a sable fur and a monocle, her hair dyed red, her make-up in screaming colors and carrying a little monkey with her. More than once, she jumps off the stage furious at a heckler shouting obscene comments to grab a bottle of champagne and smash it over his head.

Cannabis is still available in many medicinal products for various ailments, but it does not play much of a role as a psychoactive substance in the cultural life of Berlin at the time. Walter Benjman, who will also experiment with mescaline and opium, has his own reasons to try hashish, as he had already written in a 1919 letter to Ernst Schoen:

"I have read Baudelaire's Paradis artificiels. It is an extremely reticent, unoriented attempt to monitor the "psychological" phenomena that manifest themselves in hashish or opium intoxication for what they have to teach us philosophically. It will be necessary to repeat this attempt independently of this book."[105]

Walter Bendix Schoenflies Benjamin, 1892-1940

Benjamin's overall verdict on Baudelaire's hashish writings is not as negative as it would seem in this statement. He will later write in a protocol on his hashish experiences that he takes Baudelaire's writings on hashish to be the best he has encountered so far. Baudelaire's experiences with hashish definitely served Benjamin as an excellent starting point for his investigations. In 1927, the now legendary German philosopher, essayist and literary critic gets his chance to start his own investigation. When he sits down for scientific experiments on hashish intoxication with his friends, the medical doctors Ernst Joël and Fritz Fränkel, all three men take these experiments very seriously. Benjamin has a profound philosophical interest in the experience of the altered state of consciousness – he is looking for what he later calls a *profane illumination,* a new way of experiencing reality, which can lead the way to new philosophical insights. Joël and Fränkel, on the other hand, are among the group of leading experts on narcotics in the Weimar Republic, who had started a treatment clinic for morphine, cocaine and other addictions in Berlin-Kreuzberg. In his excellent essay "From 'Rausch' to Rebel-

[105] In: Benjamin, Walter (2006), *On Hashish*, Edited by Howard Eiland, „From the Letters", The Belknap Press of Harvard University Press, Cambridge, Massachusetts, p.144.

lion: Walter Benjamin's On Hashish and the Aesthetic Dimension of Prohibitionist Realism," Scott J. Thompson reminds us of the historical context in which Benjamin's experiments took place:

"Of the hundreds of books, articles, essays, monographs and dissertations on Benjamin (over 3000 exist), only a handful discuss the writings on hashish and opium (...) and none of them situate the experiments within a historical context. When Benjamin became a "test subject", he also became part of a long-forgotten community, the Weimar Republic's psychonautic avant-garde (...). The year Benjamin began his experiments (1927) Louis Lewin published his second edition of "Phantastica" in Berlin, which appears on the list of books which Benjamin read from cover to cover. (...) Hermann Schweppenhaeuser's claim that Benjamin's writings on hashish, opium and mescaline are amongst the most genuine ever put to paper can only be evaluated against the context of Weimar experimentation with psychopharmaca. Kurt Behringer's amazing monograph on mescaline, "Der Meskalin Rausch" was also published in 1927, and remains the greatest work ever written on the subject. Behringer's book contains over 200 pages of protocolls from 60 experiments in Heidelberg among doctors, medical students, natural scientists, and philosophers (...)" [106]

A path strewn with roses?

During his first experiment with hashish in December 1927, Benjamin notes in his last entry of his first short experiential protocol: *"Your thinking follows the same paths as usual, but they seem strewn with roses"*[107]

It's a neat little statement - short, picturesque and flowery, excellent for a quotation databank and, of course, several decades later in the age of mechanical, or, even more so, digital reproduction, the quote goes viral. Many commentators on Walter Benjamin's writings on hashish will use this statement to characterize and discredit the importance of Benjamin's hashish ex-

---

[106] Thompson, Scott J. (2014) "From 'Rausch' to Rebellion: Walter Benjamin's On Hashish and the Aesthetic Dimension of Prohibitionist Realism", http://www.wbenjamin.org/rausch.html, 2014.

[107] Benjamin, Walter (1927/1972), *Über Haschisch,* Suhrkamp Verlag, Frankfurt, p.68.

periments. Yet, although correctly quoted from Benjamin's first protocol on his hashish experiences, we will see that it definitely does not summarize what Benjamin had observed about the marijuana high. Nothing could be further from the truth. In fact, if we take a closer look at Benjamin's protocols, we find that Benjamin had noted and meticulously described some of the most interesting and complex thought-alterations of the cannabis high, and that his experiences with hashish profoundly influenced and inspired his thinking about other subjects.

Why, then, has his statement above been quoted so often when it comes to Benjamin's thoughts about the hashish high? It seems obvious to me that many interpreters of Benjamin have fallen prey to a biased view of the cannabis high as a merely euphoric state of consciousness with no significant useful changes in thought and cognition. Yet, why would such an outstanding and inventive thinker with so many groundbreaking ideas be so interested in hashish? Why would Benjamin plan to write book about his hashish experiences, if he experienced it merely as a euphoriant?

In what follows, I will argue on the basis of my research on the cognitive effects of a cannabis high that Benjamin's experiential protocols from his hashish experiments - which took place in Berlin, Marseilles and Ibiza between 1927 and 1934 - may require some interpretation and analysis, but are a highly interesting source for a deeper understanding of the cannabis high. Furthermore, we will see that Benjamin's cannabis experiences had a deep and positive impact on his thinking - and with it, the thinking of many other important intellectuals, not to mention generations of students in various academic fields and, finally, on society as a whole.

Benjamin's Life and Influence

Walter Bendix Schönflies Benjamin was born in 1892 in Berlin. After studying philosophy in Freiburg, Munich, Berlin and Bern, he originally pursued an academic career as a philosopher, but his unorthodox *Habilitationsschrift* – a thesis needed in Germany to qualify for professorship – fell between the disciplines of literary criticism and philosophy. Although full of brilliant insights, it wasn't well received. Benjamin declined from his request for academic promotion before receiving an expected official rejection. The text, however, would later become a classic of 20th century literary criticism. He went on to work as a freelance journalist, literary critic and essayist, barely surviving on a small subsidy granted from the Horkheimer's Institute of Social Research. He also worked for radio and translated Balzac, Proust and Baudelaire, of whom it was said he admired greatly. He was in contact and corresponding with the influential sociologists Theodor Adorno and Max Horkheimer. Benjamin was also a friend of the philosopher Ernst Bloch, who participated in his hashish experiments in Paris, and a friend of the writer Bertold Brecht.

Hannah Arendt on a German stamp

The Nazi regime's rise to power in 1933 forced Benjamin, who came from a Jewish family, to flee Germany to Paris. In exile, he met and corresponded with the philosopher Hanna Arendt, who also helped him financially. She would later describe him as one of "*the unclassifiable ones (...) those whose work neither fits the existing order nor introduces a new genre that lends itself to future classification.*"[108]

In an article "T*he Philosopher Stoned. What drugs taught Walter Benjamin*"[109], Adam Kirsch from the New Yorker aptly called Benjamin,

"*one of the central figures in the history of modernism. Benjamin approached every genre as a kind of laboratory for his ongoing investigations into language, philosophy, and art, and his ideas on the subject are so original, and so radical in their implications, that they remain profoundly challenging today (...)*"[110].

---

[108] In: Benjamin, Walter (1969) *Illuminations.* Edited and with an Introduction by Hannah Arendt. Translated by Harry Zohn. Schocken Books

[109] Kirsch, Adam (2006), "The Philosopher Stoned. What drugs taught Walter Benjamin", *The New Yorker*, August 21.

[110] *Ibid.*

Benjamin's most famous essay "*The Work of Art in the Age of Mechanical Reproduction*" (1935) left its eternal mark on our thinking about mass media and the modern art. The *Stanford Encyclopedia of Philosophy* states on Benjamin's writings:

"They were a decisive influence upon Theodor W. Adorno's conception of philosophy's actuality or adequacy to the present. (...) In the 1930s, Benjamin's efforts to develop a politically oriented, materialist aesthetic theory proved an important stimulus for both the Frankfurt School of Critical Theory and the Marxist poet and dramatist Bertold Brecht."[111]

Marijuana Insights – Myth or Fact?

Benjamin's thoughts and ideas had a strong influence on other thinkers of his time and on generations of academics, students and other intellectuals ever since. How much did Benjamin learn from his hashish experiments? Were they just eccentric excursions of a brilliant mind, leaving us with experiential protocols of largely unaltered happy thoughts, following paths "strewn with roses"? Really? In his article, Kirsch values the importance of Benjamin's work in general, but concludes that ultimately, Benjamin's drug experiments were a failure – and expresses what the view of many when it comes to Benjamin's writing on hashish:

"But what Benjamin called 'the great hope, desire, yearning to reach – in a state of intoxication – the new, the untouched' remained elusive. When the effects of the drugs wore off, so did the feeling of "having suddenly penetrated, with their help, that most hidden, generally most inaccessible world of surfaces." All that remained was the cryptic comments and gestures recorded in the protocols, the ludicrous corpses of what had seemed vital insights."[112]

Kirsch's judgment seems to be based on the widespread assumption that

[111] Osborne, Peter and Charles, Matthew (2013), "Walter Benjamin" *The Stanford Encyclopedia of Philosophy*, Edward N. Zalta (ed.), http://plato.stanford.edu/archives/win2013/entries/benjamin/.

[112] *Ibid.*

consumers of hashish or other psychoactive substances often have the feeling of great insight under the influence of the substance, only to later find that their big idea was purely nonsense or nothing special after all. Critics usually refer to this phenomenon as the "myth of insight" during a cannabis high or during the influence of other kinds of psychoactive substances. I have extensively argued before that a marijuana high can indeed cause a whole range of cognitive alterations leading to deep and valuable insights.[113] In the following, I'll argue on the basis of my research that Kirsch's negative conclusion about Benjamin's writing on hashish is wrong – dramatically wrong. Benjamin left us incredible perceptive and important observations of the hashish high and his later writing seems to have profited a lot from his experiences with cannabis.

Kirsch is certainly right to point out that the language and thinking of Benjamin's high protocols is difficult to understand – Benjamin's writing is already difficult to read when composed in a sober state of mind. Many of his hashish experience protocols have been written at least partially under the strong influence of presumably high doses of hashish, making it even harder to follow his thoughts. They are often jumpy, fragmentary, almost lyrical in some instances, and often 'cryptic'. Also, Benjamin was courageous enough not to edit out some funny and almost nonsensical thoughts during his hashish experience. It is easy to pick these out and to make fun of his writing on his hashish experiences – as some commentators seem to have done. The much more interesting work, however, is to analyze the deeper insights and observations in Benjamin's protocols and writings about the hashish experience. So, let's get down to work and look into the high mind of one of the most brilliant thinkers of modernism.

Long Gone Memories, Face Recognition, and Insights on Rembrandt

In his essay "Hashish in Marseilles"[114], Benjamin quotes his friends Ernst Joël and Fritz Fränkel, who observed and described several effects of a hash-

[113] Compare Marincolo, Sebastián (2010) *High. Insights on Marijuana,* Dog Ear Publishing, Indianapolis, and Marincolo, Sebastián (2013), *Das positive Potential von Marijuana,* Tropen Verlag, Stuttgart.

[114] In: Benjamin, Walter (1972), *Über Haschisch,* Suhrkamp Verlag, Frankfurt

ish high, including enhanced episodic memory, insights, a change in space perception and the intensification of color vision, as well as short-term memory disruptions:

"Images and series of images, long gone memories re-appear, whole scenarios and situations become present (...) he (man) comes to experiences which come near to insights and epiphanies (...) the room can become extended (...) colors become brighter, shining; (...) often, streams of thought become difficult to follow because you forget about everything you had just thought about"[115]

Benjamin himself notes how he becomes much more sensitive during his high: "*(y)ou become so sensitive: fearing a shadow would damage the paper on which it is falling*", and how his sense of space and time changes:

"The claims of space and time of the hashish eater now come to bear; and they are regal, as is well known. Versailles is not big enough for whom has eaten hashish, and eternity does not last too long."[116]

He also observes that during a hashish high he would strongly focus on faces:

"It (hashish) turned me into a physiognomist, at any rate an observer of physiognomies, and I observed something quite unique in my experience: I became dead set on the faces around me, some of them of a remarkable rawness and ugliness."[117]

These mind alterations, Benjamin writes, allowed him a deeper understanding of art: *"I suddenly understood, how a painter – didn't it happen to Rembrandt and many others? - would find ugliness appearing as a true reservoir of beauty, or better, as a treasure keeper for beauty, as the torn mountains with all of the contained gold of beauty, with beauty flashing from the wrinkles, glances, and expressions."*[118]

---

[115] *Ibid.*, p.45.
[116] *Ibid.*, p.46.
[117] *Ibid.*, p.48.
[118] *Ibid.*, p.48.

These observations are remarkable not only because they suggest that Benjamin had valuable insights under the influence of hashish, which he would later use as an art critic. They are also interesting in the face of countless other reports of marijuana users reporting in detail how marijuana helps them to better empathically understand others - which of course includes the ability to read and understand facial expressions.

Rembrandt's self portrait as Apostle Paul, circa 1661

In recent philosophy of mind, simulation theorists of human understanding have argued that our empathic understanding fundamentally relies on our capacity to imaginatively simulate the situation of others; to put us in their shoes. There are many reports from cannabis users who claim they feel this during a high, the ability to simulate others becomes enhanced. Likewise, Benjamin writes in one of his notebooks about a hashish high[119]:

"Fundamental to this feeling of empathy ["Einfühlung"] is the insinuation of one's own ego into an alien object. (...) nothing more of the person remains than an unlimited capacity, and often an unlimited propensity, for entering into the situation of every other in the cosmos, including every animal, every inanimate object."[120]

[119] Benjamin uses the German word "Rausch", which has various different translations in English, like "rush", "inebriation", "ecstasy", or "intoxication".

[120] Benjamin, Walter (2006), *On Hashish*, From the Notebooks, Preparatory sketches for "On Some Motifs in Baudelaire", edited by Eiland, Howard. The Belknap Press of Harvard University Press, Cambridge, Massachusetts, p.142.

Connectedly, many marijuana and hashish users have remarked that during a high, they are not only able to understand other people better, but also music or art, even those forms of which would not normally resonate with them. Lester Grinspoon once told me in a private conversation that marijuana helped him to expand his musical taste spectrum and to enjoy listening to the music of the Beatles, although he preferred classical musical before. Likewise, Benjamin notes in a protocol from a hashish experiment:

"Feeling that I understand Poe much better now. The doors to the world of the grotesque seem to be opening. (...)" [121]

Hyperfocusing, Mind-Racing, and Humor

Benjamin's strong attentional focus on faces during one of his hashish experiments is only one instance of the "hyperfocus" – effects of attention reported by him. Later in his essay, Benjamin reports another observation made possible by this hyperfocusing: *"There were times in which the intensity of acoustic impressions made them supersede everything else."*[122]

During his high, this strong attentional focus helps him to understand what a strong dialect he is hearing in conversations around him:

"The most peculiar thing about this noise coming from voices was that it did sound completely like a dialect. People from Marseille as it were didn't speak French well enough for me."[123]

I have argued previously that one of the interesting effects of a cannabis high is an acceleration of associative thought, or during a stronger high, an accelerated stream of visualized images. Benjamin also observes this effect of acceleration when he tries to trace back Baudelaire's inspiration for his poem "Les Sept Vieillards" to the use of hashish:

[121] *Ibid.*, "Hauptzüge der ersten Haschisch Impression", p.66.
[122] Benjamin, Walter (1972), *Über Haschisch*, Suhrkamp Verlag, Frankfurt, p.53.
[123] *Ibid.*, p.54.

"Here, human reason becomes more flotsam, at the mercy of all currents, and the train of thoughts is ***infinitely*** *more accelerated and 'rhapsodic."*[124]

In another instance, he also describes a "stormy production of images" during a hashish high: *"About the images themselves I cannot really say much here, because of the tremendous speed with which they arose and then vanished again; (...)"* [125]

Benjamin also observes that during his high, he experienced a wonderful, inspired humor - and he proves this claim in his protocols with statements written under the influence that could have come from comedy genius Groucho Marx: *"If Freud would psychoanalyze God's creation the Fjords wouldn't come off very well"*[126]

Another one of his funny experiences during a high is about *Pâté de Lyon* (duck liver paste from the French city of Lyon):

"Lion pâté, *I thought to myself laughing funnily, as it lay in front of me on a plate, and then, despicably: this rabbit or chicken pâté – whatever it may be. I was hungry like a lion, so it seemed not inappropriate to me to satiate my hunger with a lion."*[127]

As nonsensical as this might seem, Benjamin's observation about his funny association is also interesting. Under normal circumstances we would not think of "Lyon" or "*pâté* de Lyon" as having anything to do with a lion. The association of Lyon and Lion is somewhat obvious but usually we use many expressions and names in our discourse without thinking about their metaphorical content. Names and many expressions usually become 'opaque' to us in our everyday use. Benjamin's reflections on his associations during a high connected to the expression *pâté* de Lyon seem funny, nonsen-

[124] Benjamin, Walter (2006), *On Hashish*, From the Arcades Project (1927-1940) Edited by Howard Eiland, The Belknap Press of Harvard University Press, Cambridge, Massachusetts, p.138.

[125] Benjamin, Walter (1972), *Über Haschisch*, Suhrkamp Verlag, Frankfurt, p.108.

[126] *Ibid.*, p.120.

[127] *Ibid.*, p.49.

sical and not really useful at this point - but it is clear that the process of changing one's perception on language during a high can help thinkers to come to a deeper understanding of language and to interesting associations along the lines of its underlying metaphorical content.

Let me summarize then: Benjamin and his friends experienced and described many of the cognitive enhancements of a cannabis high which I have researched and discussed in my books and essays. We find wonderful descriptions of the effect of the hyperfocusing of attention, of an enhanced episodic memory, changes in the perception of space and time, insights, an enhanced pattern recognition (seeing for instance new patterns in faces or in art) and an enhanced capacity for empathic understanding. Clearly, even if Benjamin's protocols make liberal use of poetic language and are often hard to read, there is much to find if you are interested in the ways in which a cannabis high influences cognitive processes. Benjamin's elaborate characterizations of altered thinking during a hashish high show that his early statement *"(...) your thinking follows the same paths as usual, but they seem strewn with roses"* does not reflect his overall thinking on the subject.

I have pointed out above that Benjamin's hashish experiments did not only produce insights about the effects of hashish, but had a lasting influence on his perception and thinking of art. In what follows I will show how deep Benjamin's understanding of the marijuana high was – how insightful some of his statements really are in the light of my research – and how much these experiences during a high really informed one of the most influential thinkers of modernism.

Gestalt Psychology and the Functional Shift in Perception

In his second protocol about his hashish experiences written in 1927, Benjamin explains an expression he said he took from his friend, Ernst Joël:[128]

---

[128] Ernst Joël had been treated with the painkiller morphine after being wounded in the First World War. After the war, Joël and his friend Fritz Fränkel started a clinic for addicts in Berlin. Later they would start to experiment with psychoactive substances and

"Functional shifting. I take this expression from Joël. Here is the experience that led me to it: In my satanic phase someone gave me one of Kafka's books: "Betrachtung." I read the title. But then the book immediately became to me what a book in the hand of a poet becomes to the academic sculptor with the task to make a statue of this poet. It was at once integrated into the sculptural form of my body (...)" [129]

So, yes, again, this sounds cryptic. But this paragraph is definitely worth a second look, because it contains a key to Benjamin's insights on marijuana. Being high, Benjamin at first perceives the book as a book under its everyday function: he looks at the object in his hands as a book, reads the cover. But then his perception shifts away from this function and he suddenly sees the object like an artist would look at it, as somebody who only wants to chisel its form out of stone.

Benjamin described in one of his high protocols one of the most important psychological mechanisms fundamental to the process of spontaneous, creative insight – one of the mechanisms often triggered during a high, which I believe helps to explain why so many users have reported valuable insights during a high. Let me explain.

While Benjamin and his friends experimented with hashish in Berlin, a group of psychologists led by Max Wertheimer – also located in Berlin - were working hard on their groundbreaking theory about human perception which would later become famous under the name *Gestalt* psychology ("*Gestaltpsychologie*") or gestaltism. One of their main goals was to observe and explain the phenomenon of creative insights in thinking. But insights seem so elusive in our everyday lives; they often seem to come out of nowhere. We do not really know how to generate them. If you ask somebody how he came up with a certain creative insight, he will often be unable to provide an explanation. Obviously, then, the processes involved are mostly unconscious, and Wertheimer was convinced that there was something special about this kind of "productive thinking" that culminates in a so-called *Eureka* moment. Wertheimer and his followers wanted to study "productive thinking"

---

initiated Benjamin to participate in experiments with hashish.

[129] Benjamin, Walter (1972), *Über Haschisch*, Suhrkamp Verlag, Frankfurt, p.75.

empirically so he started to design intellectual problems that required a certain non-linear creative solution, an insight from the problem solver. These problems were then given to test subjects in an experimental setting so that psychologists could study how the process of insights actually works.

Some years after Benjamin's observations about the "functional shift" in his perception during a high, Karl Duncker, Wertheimer's most talented student, came up with his famous candle problem (sometimes called 'matchbox problem'), a problem designed to be solved only by the means of a creative insight.[130] The setup is simple: test subjects were given a matchbox, matches, a candle and some thumbtacks.

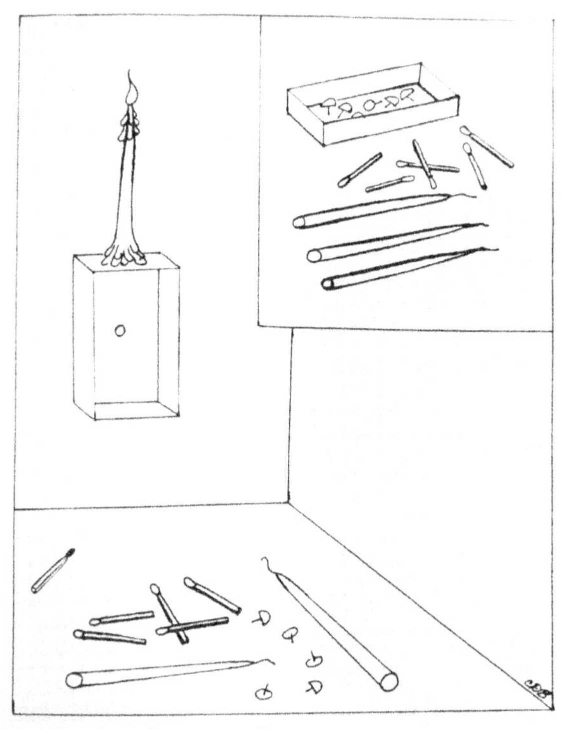

Karl Duncker's famous candle experiment

The task given by Duncker to his test subjects was to fix the candle to the wall using only the objects they were given. Note that the candle can not be fixed directly to the wall – the creative insight needed here was to see the matchbox container as a tray for the candle, fix it to the wall and put the candle on it.

Duncker showed that subjects needed longer to solve the problem if he presented them the matchbox with the matches inside, instead of presenting the matches separately – thereby emphasizing the function of the matchbox as a container for matches. His hypothesis was that the perception of test subjects was '*functionally fixed*'

[130] Duncker, Karl (1935). *Zur Psychologie des produktiven Denkens*, [*Psychology of Productive Thinking*]. Springer. OCLC 6677283.

to seeing the box as a container – which would prevent the insight to use it as a tray. Only if the subjects moved away from this perception, would they be able to arrive at the insight that would solve the problem.

Duncker's candle experiment and his concept of 'functional fixedness' has become famous and modern theories of insights have clearly confirmed that Duncker's notion was a groundbreaking step in characterizing one of the most important mechanisms in the processes of insights in the phase that leads to the *Eureka* moment.

Now, back to Benjamin's observation. It should be easy now to see the importance of Benjamin's description of what he and Joël called a "functional shift" in perception during a high. Years before the *Gestalt* psychologists would come up with the notion of "functional fixedness", Benjamin had given an explicit description of one of the fundamental keys to understanding why – as he and his friends had observed themselves – a cannabis user would sometimes experience important insights and have *epiphanies* during a high. Benjamin had observed that during a high, there can be a functional shift in one's perception of an object as had happened for him with Kafka's book; he was not bound anymore in his thinking to seeing the object merely under the function of a book – and seeing objects with this more open perception generally enables subjects to come to interesting new ideas how to use those objects in a new way. In his early hashish experiment, then, Benjamin had named and described an important cognitive mechanism that can lead to insights, as *Gestalt* psychologists would later demonstrate.

What Hashish Did to Walter Benjamin

Benjamin's experiments with hashish were a success. He never really succeeded in writing the book on hashish or drugs he wanted to write because he had to flee from the Nazi regime. But the *posthumous* compilation of his essays in the book *On Hashish* and other passages on hashish in his wider writings contain several outstanding observations on many interesting cognitive alterations of the cannabis high. Benjamin also showed us how these effects on higher cognition during a high can be used positively: for facial pattern recognition, to revive long gone memories, for a deeper appreciation of art and nature, to hyperfocus on certain perceptions and actions, for a broader sense of humor, for an enhancement of one's empathic ability to take the perspective of others, and for insights. As to the latter, Benjamin even made observations about a "functional shift" in perception during a high, which helps to explain why so many cannabis users have reported an enhanced ability to generate insights during a high.

We know that Benjamin intended to use the highs of his hashish experiments to think about philosophical and other intellectual themes. So, how much did his experiments influence the rest of his important work? Did hashish help Benjamin to develop new ideas, generally? In his first protocol about a hashish high from 1927 he states:

"17. It seemed to me: a marked unwillingness to converse about matters of practical life, about the future, data, or politics. One seems to be drawn to the intellectual sphere just like some maniacs are drawn to the sexual sphere."[131]

In the beginning of March 1930, Benjamin reports about a high that the most important part was that with his eyes closed he had a thought about the nature of what he calls "*aura*".[132] This is the first time Benjamin mentions this notion; it will play a fundamental role in his later work. In this first mentioning, he explains that for him, the *aura* is not some esoteric mystic property, but something real that can be assigned to each and every object – not just some special objects. Benjamin believes the *aura* changes when the object moves and calls it an "*ornamental environment ("Umzirkung"), in which the object or person is enclosed.*"[133] It is certainly difficult to find a simple definition of Benjamin's notion of *aura* – we can find various different characterizations throughout his work. For now, it should suffice for an initial understanding that Benjamin thought of it as the concrete object in its unique spatio-temporal and cultural-historical context. The concept of *aura* will play a fundamental role in his thinking about art in his most famous and widely influential 1936 essay "The Work of Art in the Age

---

[131] Benjamin, Walter (1972), *Über Haschisch*, Suhrkamp Verlag, Frankfurt, p. 67.
[132] *Ibid.*, „Haschisch Anfang März 1930", p.106.
[133] *Ibid.*

of Mechanical Reproduction."

Clearly, then, Benjamin thought about some themes of his later work under the influence of hashish with the intent to come to new insights. Now, of course, the question is whether the hashish high really helped him to gain important insights, as he himself had stated in his writings. Or was he just a brilliant mind who came up with great ideas, anyway – some of them during a high? I will argue in what follows that we can actually trust Benjamin's own judgment: his experiences with hashish actually helped him to gain insights which inspired his later work.

Let us look at two examples of Benjamin's ideas in his famous essay "The Work of Art in the Age of Mechanical Reproduction" to argue this point.

(1) The 'Functional Change' of Art in the Age of Mechanical Reproduction

In this groundbreaking essay, Benjamin writes that the first pieces of art had their ultimate foundation in the magical or religious rituals in which they were used – think, for example, of ancient statues of gods in temples used in various religious rituals. This ritualistic or religious function ('utility value') is part of what he calls their *aura*, the cultural, historical, scenic embedding, which makes each of them unique. Even later in times of the Renaissance, Benjamin proceeds, art goes on to serve a certain function, and is then fundamentally embedded not in religious rituals, but in secular rituals, serving an aesthetic ideal. One of the core ideas of Benjamin's groundbreaking essay is to observe that modern means of technical reproduction strip pieces of art of their *aura.* A photograph of an object can be reproduced and seen anywhere in the world, taking the object out of its spatio-temporal and historical-cultural context. Thus, *"(...) for the first time in world history, mechanical reproduction emancipates the work of art from its parasitic dependence as ritual."* [134]

During a hashish high, Benjamin experienced and described that objects like that of Kafka's book can undergo 'functional shifting' in our perception; they are no longer perceived under their original function of everyday use.

[134] Benjamin, Walter (1936), „The Work of Art in The Age of Mechanical Reproduction", in: Morra, Joanne, and Smith, Marquard (eds.) (2006) *Visual Culture: Experiences in visual culture,* Taylor and Francis, *p.19.*

Now, Benjamin explains that through the use of technical reproduction such as photography or film, artifacts are taken out of their original context of function. This process, also, constitutes a functional shift: a piece of art is not founded anymore in its function within a ritual, but it becomes founded in "politics" – now, what makes a piece of art valuable is not its function within a defined ritual, but the question whether it can be exhibited in certain environments. Benjamin does not use the description 'functional shift' at this point, but he uses the almost synonymous 'change in function': *"When the age of mechanical reproduction separated art from its basis in cult, the semblance of its autonomy disappeared forever. The resulting change in the function of art transcended the perspective of the century; for a long time it even escaped that of the twentieth century (...)."*[135]

Benjamin's experiences with hashish made him see objects in a new way; the high caused what he then called 'functional shift'. In his high experience, he "stripped" Kafka's book of its original *aura*, of its function as a book, to see it merely as an arbitrary object that could be chiseled out of stone; it seems at least plausible that this experience helped him to see how art reproduced through technology – especially photography and film – led to a similar functional change; the objects presented in film are stripped of their *aura*. The medium of photography or film allows us to see objects out of context, leading to a functional change in our perception.

And when Benjamin goes on to explain this idea in more detail in connection with the medium of film, he presents some more fundamental insights which, again, seem to be linked to his previous experiences with hashish:

(2) Film Technology allows a Revolutionary New Experience of Reality

For Benjamin, there is a revolutionary potential in this functional change of art. For him, especially the medium of film shows a complete change in the function and the potential of art; with its technical means, it offers a whole new experience not only to a select group of eclectic viewers, but to the masses around the world. Benjamin was excited how the technical means

---

[135] *Ibid.*, p.127.

of film allowed for so many of us to experience a deeper understanding of the world:

"*For the entire spectrum of optical, and now also acoustical, perception the film has brought about a similar deepening of apperception.*"[136]

He goes on to explain exactly in which ways the technical means of film can achieve this by offering us what we could call new 'modes of presentation' of reality:

"*With the close-up, space expands; with slow motion, movement is extended. The enlargement of a snapshot does not simply render more precise what in any case was visible, though unclear: it reveals entirely new structural formations of the subject. So, too, slow motion not only presents familiar qualities of movement but reveals in them entirely unknown ones "which, far from looking like retarded rapid movements, give the effect of singularly gliding, floating, supernatural motions.*"[137]

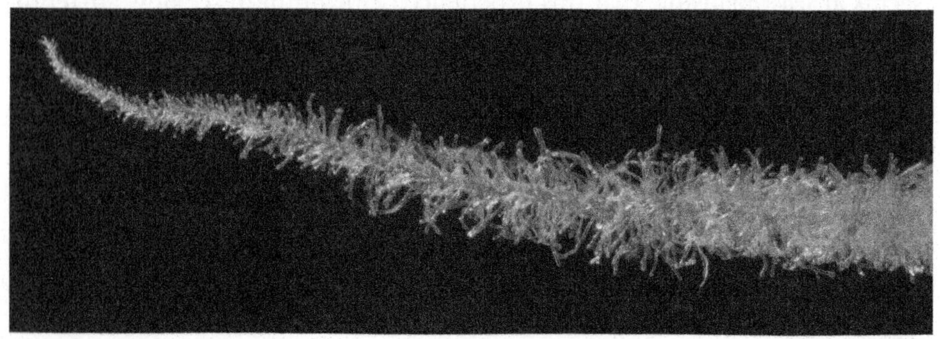

An extreme close-up macro photo of a cannabis stigma, a part of the flowers. Photo © Sebastián Marincolo 2012

Benjamin goes on to explain that the camera can show us, for instance, not only the usual visibly movement of somebody walking, but also various postures during that movement that only come out in a fraction of a moment, postures that we would never see with the naked eye. The techniques of filming can help us stretch our experience to see aspects of reality of which we were previously unaware, and can therefore act like the psychoana-

[136] *Ibid.*, p.127.
[137] *Ibid.*, p.127.

lyst who helps his patient to become aware of his unconscious thoughts: *"Here the camera intervenes with the resources of its lowerings and liftings, its interruptions and isolations, it extensions and accelerations, its enlargements and reductions. The camera introduces us to unconscious optics as does psychoanalysis to unconscious impulses."* [138]

These changes of experience, these altered modes of presentation of reality sound familiar when we look at Benjamin's observations during his hashish experiences. In his protocol from a high in Marseilles in 1928 he notes, for instance, how he intensely focuses his attention on the handle of a coffee pot, just like in a close-up in a film: *"The handle of a pot, with which coffee is served here, starts looking very big, and stays like this"* [139] As I have explained before, the effect of "hyperfocusing" is a quite fundamental effect during a cannabis high, so it is not surprising that Benjamin notices this change in perception.

As quoted above, Benjamin had also noted and described 'mind racing' during a cannabis high, an accelerated stream of thoughts or images, and the enhancement of episodic memories. In my book, *High. Insights on Marijuana*, I described how this acceleration of thought and imagery of memories can lead to a revelatory mode of presentation which can be compared to the technique of time-lapse in film. The groundbreaking experimental movie Koyaanisqatsi, directed by Godfrey Reggio, made explicit use of techniques like time-lapse to communicate some basic insights:

"The mechanical and natural patterns shown in Koyaanisqatsi are omnipresent in our daily life: we often experience traffic in cars, and we see the clouds moving in the sky. But only when we see these patterns in a more compressed mode of presentation do we start to attend to them as such; usually, when I look at the traffic passing through several synchronized traffic lights, I am not interested in the fact that they "dance a mechanical rhythm", since I am momentarily just interested in crossing the street or getting home in my own car. The mode of presentation in the movie Koyaanisqatsi, which shows many non-commented time-lapse footage, focuses our attention on the very rhythm of our civilized mod-

---

[138] *Ibid.*, p.127.

[139] Benjamin, Walter (1972), *Über Haschisch*, Suhrkamp Verlag, Frankfurt, p.96.

ern life and nature. It show us the omnipresence of patterns we have not noticed before, even though they determine almost all of our actions and interaction in modern society as we take trains, drive cars, work in factories or companies (...)."[140]

During a high, you can have a similar "time-lapse" effect. The changed mode of presentation of memories during a high can also help you to see a new pattern in reality, an aspect of reality that was invisible to you before:

"Mind racing through past memories of your childhood and adolescence under marijuana, you may find that you have been just as stubborn in your behavior towards your parents as you are now in discussions with your wife. It is almost like seeing or having a chain of associated memories played in time-lapse."[141]

As we have seen, Benjamin experienced and described various effects of a hashish high which play a role in such a time-lapse experience; the enhancement of episodic memory, the acceleration of thought or imagery. Remarkably, these associative links during a high often present us with a chain of similar situations; it is as if you would rapidly make an associative chain between similar memories, such as seeing a rapid succession of examples of stubborn behavior throughout your life – which can lead to the insight, for instance, that you have kept a certain behavior throughout your life. In one of his experiments in 1928, Benjamin also notes that he can easily find similarities between objects in his memory during his high:

"I immersed myself in intimate contemplation of the sidewalk before me, which, through a kind of unguent (a magic unguent) which I spread over it, could have been–precisely as these very stones–also the sidewalk of Paris."[142]

The analogy between the special illuminating experience of time-lapse in film and during a high show that Benjamin's experiences with hashish could

---

[140] Marincolo, Sebastián (2010), *High. Insights on Marijuana,* Dog Ear Publishing, Indianapolis, p.107.

[141] *Ibid.*, p.108.

[142] Benjamin, Walter (2006), *On Hashish*, From the Arcades Project (1927-1940) Edited by Howard Eiland, The Belknap Press of Harvard University Press, Cambridge, Massachusetts, p.54.

have paved the way for him to understand the ways in which the medium of film and its technical means, can help to expand the experience of reality, to enable all of us to see aspects of reality that could not be seen before – just as the experience of the cannabis high can allow users to see new aspects of reality.

So, back to the original question. What did hashish do to Walter Benjamin? He had experimented with hashish with the philosophical intent to come to experiences that gave him a deeper understanding of reality. In his hashish experiments, he obviously experienced and meticulously described various interesting changes in perception and cognition which offered exactly that: an altered state of consciousness enabling new perspectives and revealing new aspects of reality. Benjamin's search for what he called a 'profane illumination' during the influence of hashish was successful. In his later work, he used the insights gained during these experiences to come to a radical new understanding of the role of film and of art, in general, in modern society – and his insights have gone on to inspire not only his influential philosophical and literary friends, but generations of students and other readers ever since.

Benjamin, "Rausch", and Revolution

Benjamin knew very well that the experience of the high ("*Rausch*") does not only have a positive mind-altering potential for the individual but also bears a revolutionary political potential – just like the medium of film. In his essay "Surrealism" (1929), he wrote:

"Lenin called religion the opiate of the masses (...) creative overcoming of religious illumination certainly does not lie in narcotics. It resides in a profane illumination, a materialistic, anthropological illumination, to which hashish, opium, or whatever else can give an introductory lesson. (...) To win the energies of intoxication for revolution – this is the project on which Surrealism focuses in all its books and enterprises."[143]

[143] Benjamin, Walter (2006), *On Hashish*, from: "Surrealism" (1929) Edited by Howard Eiland, The Belknap Press of Harvard University Press, Cambridge,

In a letter to Max Horkheimer in 1938, Benjamin states:

"Critical theory cannot fail to recognize how deeply certain powers of intoxication ["Rausch"] are bound to reason and to its struggle for liberation."[144]

As with the medium and the experience of film, Benjamin was well aware that the experience of a certain kind of "*Rausch*" had been abused by fascists and other militaristic forces driven by capitalist greed. The threat of fascism had long been omnipresent in his life. In 1940, the year in which Benjamin tries to flee from Hitler's influence to neutral Spain, German soldiers invade Paris in a "Blitzkrieg", fueled by millions of pills of *Pervitin,* also called "*Hermann-Göring-Pillen*", methamphetamine pills ("Crystal Meth"), which helps them to stay awake, keeps them fit and focused for days, dampens their fears and empathy. Hitler, himself a multiple-toxico-maniac, receives high dosed shots of vitamins, crystal meth, "Eukodal", which is a derivative of morphine, cocaine applications for his head aches, and many other substances until his death.

Portbou, Spain

Benjamin decided shortly before the invasion of German troops in France to flee to Spain to then proceed to Portugal and eventually to the U.S. After a strenuous climb over the Pyrenees Benjamin, who suffers from both heart and lung disease, makes it to Portbou in Spain in September 1940 with a small group of refugees only to be denied asylum. The group struggles not to be deported back to France with no success. At the age of 48, Benjamin dies in the hotel in Portbou where the refugees are held captive. He either commits suicide using the morphine that he carries for

Massachusetts., p.132.

[144] *Ibid.*, "From the Letters", p.145.

that purpose in the eventuality that he is taken captive, or he is killed by the Gestapo or by Stalin's agents. We will probably never know.[145]

Benjamin's friend, the philosopher and neo-Marxist Ernst Bloch, who had experimented with hashish with Benjamin in Berlin, also went on to influence a whole generation of students in the 1960s with his main work *The Principle of Hope*, which he wrote in the U.S. while in exile between 1938 and 1947. In his book, Bloch mentions his hashish experiences and how they can help to enhance a user's ability for imagination – a gift that for Bloch was central to human nature, enabling and driving us to live towards a utopian society.

More than three decades before the revolutionary movements of the late '60s in various countries around the world, before the music at Woodstock stunned and opened the minds of hundreds of thousands of people, young and old, under the influence of marijuana or LSD, Benjamin wrote in his essay "One-Way Street":

"The ancient's intercourse with the cosmos had been different: the ecstatic trance ["Rausch"]. For it is in the experience alone that we assure ourselves of what is nearest to us and what is remotest from us, and never of one without the other. This means, however, that man can be in ecstatic contact with the cosmos only communally. It is the dangerous error of modern men to regard this experience as unimportant and avoidable, and to consign it to the individual as the poetic rapture of starry nights. [146]

---

[145] For an interesting report on Benjamin's final hours, see "Chronicling Walter Benjamin's final hours", *HAARETZ*, Nov. 2014, *http://www.haaretz.com/chronicling-walter-benjamin-s-final-hours-1.449897*

[146] *Ibid.*, p.129-30.

The Most Powerful Drug Used by Mankind

Your mood instantly changes when your lover calls you her "hero"; the verdict of your elementary school teacher that "you don't have talent for anything" can leave scars in your mind for decades and Bob Marley's demand to "stand up for your rights" may change your mind and guide your actions forever after. Words are a powerfully mind-altering 'substance'. They can change and influence our minds every day, rapidly, profoundly and enduringly. As the writer and poet Rudyard Kipling once said, "*words are, of course, the most powerful drug used by mankind.*"

Rudyard Kipling (1865-1939), Photo by Joseph John Elliott and Clarence Edmund Fry

Obviously, when it comes to the discussions over psychoactive substances, words, slogans and their underlying metaphors have long shaped our thinking. We tend to underestimate, however, how perfidiously certain expressions and slogans have been designed to influence our minds and to narrow down our perception.

Don't swallow the bait!

The American linguist George Lakoff reminds us that generally, our thinking is deeply influenced by metaphors that can be found in our concepts, slogans and expressions. In his book, *Don't think of an elephant! Know Your Values and Frame The De-*

bate"[147], Lakoff describes how the conservatives in the U.S. have managed to 'frame' debates with expressions in a way that makes it hard to argue with them. One of his main examples is the expression "tax relief" which came out of the White House right after President George W. Bush took office. Lakoff's point about the metaphor behind the expression is that for there to be a relief there must be some party who *afflicts* people, as well as a *reliever,* who then is the *hero.* Lakoff thinks that it was an important mistake of the Democrats to take up the expression, and, thereby, to affirm the underlying metaphor and to discuss the issue under the label "tax relief". By using this expression they actually endorsed a worldview as framed by the Conservatives and thereby had already lost the debate. Even good, rational arguments against certain tax cuts would always leave them as the villains, trying to stop the "hero" *relieving* the "victims." The Democrats had swallowed the bait. Lakoff argues that the Democrats should have rather endorsed a different language that better expressed their world view, namely, a metaphor that would express that taxes are investments in the future made by citizens, which can turn out to be profitable by leaving them with a better infrastructure and many other important benefits.

"Drugs" and "Wars"

Lakoff's analysis is an important reminder when it comes to discussing psychoactive substances and their role in society. The most prominent example of a slogan from the Republican party that framed these debates and is still used by many is the "War on Drugs", which was introduced by the Nixon administration in the 1970s. There are so many things wrong with this expression that I hardly know where to begin, so let me just pick out a few important aspects here. The expression "War on Drugs" implies that there is something intrinsically wrong about all drugs and drug use in general, which makes no sense. Obviously, this is meant to be a war on certain drugs; it is not a war on alcohol, nicotine, caffeine nor is it a war against the thousands of prescription drugs freely available in today's world. You cannot wage a war against substances. It could only be a war against people using or

---

[147] Lakoff, George (2004), *Don't Think of an Elephant. Know Your Values and Frame The Debate.* Chelsea Green Publishing Company, White River Junction, Vermont.

dealing *certain* drugs, those deemed unacceptable by the powers that be.

Furthermore, it is not a classic *war* between armies; it is rather a militarily enforced repression of a certain group of people using those drugs – and, note that the use of the term "drug" is significantly imprecise here. Drugs are usually defined as natural or synthetic substances that affect our bodily functions – drugs do not necessarily have to be mind-altering. When talking about psychoactive drugs like alcohol, marijuana, nicotine, sleeping pills, antidepressants or others in a similar vein, I therefore suggest to be more precise and to use the notion "psychoactive substance".

So, really, there has never been a war on drugs, only the violent repression of certain groups of people using or dealing some types of psychoactive substances. If there is a *war* at all, it is a "War on Cognitive Liberty", a war that has been (and is still being) waged to fight those who freely choose to alter their minds with specific substances. I can only hope that people will soon become more critical about the language of the "War on Drugs" before a new Republican government comes along successfully announcing a new upgraded nonsensical slogan like a "Star Wars on Drugs", promising to use innovative satellite-based observation techniques and shiny little drones. Yes, that sounds funny. But why haven't we laughed about the notion of a "War on Drugs" in past decades? We swallowed the bait, a verbal bait that is so obviously poisoned, but we are still using it - liberals and conservatives alike.[148] Millions of people still have to pay the price for this flawed thinking. Also, as another consequence from such linguistic conditioning, hundreds of millions of alcohol consumers do not consider themselves as what they really are: drug consumers with a long history of drug consumption. They have been led to think of themselves as alcohol consumers, but certainly not drug consumers

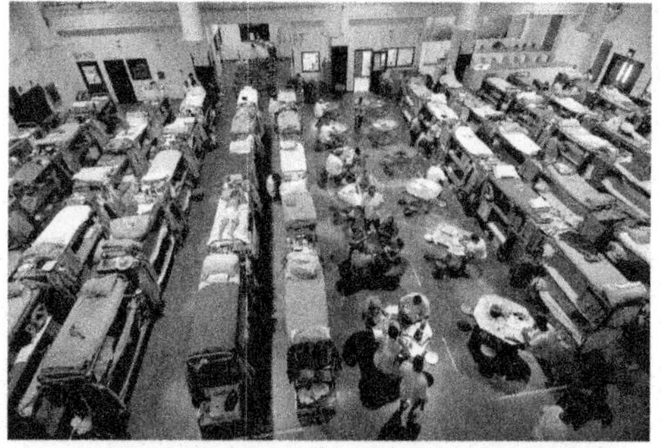

From Wikipedia commons: "The crowded living quarters of San Quentin Prison in California, in January 2006. As a result of overcrowding in the California state prison system, the United States Supreme Court ordered California to reduce its prison population."

[148] The Obama administration has officially stopped using the "War on Drugs" rhetoric; sadly, however, they keep up many of the repressive measures.

– even though they know that alcohol is a drug. Even the liberal press writes that British singer Amy Winehouse died of drugs *and* alcohol; a phrase that sets alcohol aside from the category of drugs. And this ridiculous act of self-deceit still has a marked influence when it comes to our thinking about the role of drugs in society. We know that alcohol is a drug and we know where it is sold. But we just don't see bars, supermarkets, liquor stores and restaurants as what they are: drug dealers.

Watch Your Language, Change Your Mind

These observations about language use are not only important when it comes to discussing the consumption or regulation of psychoactive substances. We are all are using countless metaphors that to a large degree have been produced by very well paid and highly talented spin doctors working for decades on campaigns against marijuana and other mind-altering substances. These metaphors have altered our minds, narrowed down and tilted our perception of psychoactive substances, of our own use and behavior.

If you are consuming a psychoactive substance, be it legal or illegal in your country at the time, you are not by definition a *drug abuser*. You are not a *drug abuser* if you *use* a cup of coffee once in a while if it helps you to be more alert while you work. And you are not a *drug abuser* if you use marijuana to get high to work creatively or to help you to focus your attention on your partner to get empathic insights. You can abuse both legal and illegal psychoactive substances, but whether your activity should be called use or abuse ultimately depends on how we evaluate the effects of the respective substance coming from your personal method of consumption.

It is now time to rethink our language, to be aware and to critically oppose the metaphoric content of slogans about psychoactive substances that have been conjured up by spin-doctors in recent decades. We need to watch our language to change our minds, to start a mind-altering process not only individually, but also for society as a whole. As with many other propaganda campaigns against marijuana in the recent past, the "War on Drugs" campaign has been used to instill fears in a wider public. Fears are an easy way to narrow down our perception and to influence and control our actions. But, to quote the ever-insightful Rudyard Kipling again, *"(...) of all the liars in the world, sometimes the worst are our own fears."*[149]

[149] Cannabis may have had an important positive influence on Kipling's life, too. Charles Allen writes in his book *Kipling Sahib: India and the Making of Rudyard Kipling* how the 18-year-old Kipling suffered a severe bout of dysentery in 1884 and smoked opium and ingested 'a stiff dose of chlorodyne' for relief: *"Dr. Collis-Browne's Chlorodyne, patented in 1871, was a mixture of opium in alcoholic solution, tincture of cannabis and chloroform. 'There is convincing evidence that this double dose hit him with the force of a revelation,' writes Allen. 'In modern parlance, it 'blew his mind,' opening the doors of his unconscious hitherto kept tight shut and causing him to lose some of his fearfulness (...) [it] brought a new dimension to his thinking (...) freeing him to speak more directly from within himself (...)."* From: *veryimportantpotheads.com, entry on Rudyard Kipling, Sept. 2013.*

An Unusual Argument for the Legalization of Marijuana

There are many compelling arguments against marijuana prohibition already out there – I will not repeat them at this point. In this essay I would like to add an unusual line of thought which constitutes another reason for the legalization of marijuana.

The Experience of Millions of Marijuana Users

According to the WHO (World Health Organization), approximately 150 million people around the world today are using cannabis for its psychoactive effects. I doubt that this estimate is correct – the actual number is probably much higher – but let's leave it at that for now. Some of these users value diverse cannabis strains for their respective mind-altering properties, whereas others choose strains suited mainly for a variety of medicinal purposes. The vast majority of people who consume cannabis have to obtain it illegally because of the prohibition laws in most countries. A great deal of the cannabis sold on the black market has been produced under bad conditions, grown purely for profit, and sold to the consumer without quality controls of any kind.

In this situation, consumers often do not know exactly what kind of cannabis they are using – an Indica or a Sativa-dominant strain, a Kush type, an "auto"-flowering strain (hybrids containing cannabis ruderalis, arguably another species of cannabis), let alone the type of the strain. They also do not know how the marijuana they buy has been produced and stored, if growers have used biological or chemical fertilizers, if they refrained from using pesticides or indeed any of its background. An illegal market does not generally allow for much transparency.

Currently, there are millions of cannabis users worldwide who have millions of anecdotes and stories to tell about the effects of marijuana – but most of them are unable to give precise details about which plants they used, or the conditions under which these plants were produced. Not only is this information virtually impossible to obtain in an illegal market, the worldwide prohibition also rests on a disinformation campaign about marijuana which has helped to generate a taboo around its use. Prohibition prevents independent, expert information on cannabis biology, strains, genetics and

Macro photography of a cannabis seed. Photo © Sebastián Marincolo 2012

growing from entering the mainstream media. Furthermore, many users have even been actively misled by criminal dealers who take advantage of prohibition and the lack of knowledge on the side of consumers to sell them cheaply produced, low-grade marijuana – sometimes laced and weighted with dangerous substances – under the name of superior strains.

The Importance of Anecdotal Evidence

As marijuana expert Lester Grinspoon points out, anecdotal evidence is a valuable source for much of our knowledge about the effects of synthetic medicines as well as plant derivatives. He also rightly reminds us that

"(t)oday, advice on the use of marijuana to treat a particular sign or symptom, whether provided or not by a physician, is based almost entirely on anecdotal evidence."[150]

Lester Grinspoon himself helped to accumulate much of the valuable anecdotal evidence upon which scientists and patients can draw today, with his websites on medical marijuana (*rxmarijuana.com)* and on the enhancement uses of marijuana *(marijuana-uses.com)*. Another website where users are asked for their experiences with marijuana (and other psychoactive substances) is *erowid.org*. This site makes available thousands of pages of information on psychoactive substances, and specifically asks users for the drug and dosage they used and for their reports on the effects observed. Naturally, as Grinspoon also reminds us, anecdotal evidence has to be examined with care and must be critically evaluated. But it leads the way, as it always has, for researchers who are interested in the effects of all kinds of substances on the human brain and body.

Dozens of Cannabinoids, Terpenes, and Flavonoids

Cannabis has been used as a medicine for thousands of years and is now used in the treatment of many medical conditions, such as chronic pain, nausea, insomnia, neurogenic pain, asthma, glaucoma, epilepsy, Tourette's Syndrome and so on – the list seems endless. We are now beginning to see that the effects of marijuana have to be understood as an interaction between various cannabinoids and other chemicals like the terpenes and the already existing endocannabinoid system in our bodies. For a long time we have known that not only the main active ingredient THC (Delta 9-

---

[150] Grinspoon, Lester (2009), "Medical Marijuana: A Note of Caution *non nocere*", *http://rxmarijuana.com/note_caution.htm*.

Tetrahydrocannabinol), but also other cannabinoids like CBN (cannabinol), CBD (cannabidiol) THCV (tetrahydrocannabivarin), and CBC (cannabichromene) play an important role in producing or modulating the psychoactive effects of marijuana. Cannabis strains contain not only more than 100 cannabinoids, but also around 200 terpenes and more than 20 flavonoids; some of these chemicals are already known to have psychoactive properties, others are known to be medically effective in a variety of ways.

There are millions of marijuana users around the world consuming hundreds of different cannabis strains produced by various seed banks and breeders today. These people are not simply using marijuana to induce relaxation or some kind of euphoric state, as uninformed non-users frequently assume. They value the diverse effects of marijuana and many are using it consciously and deliberately to alter their mind for very specific activities. Some want to be able to focus better on their work, others use it to remember the distant past; to make or listen to music; to see new patterns and to come up with new ideas; to get emotionally in touch with themselves or others; or to make love with their partners. Others are legally using natural strains for dozens of various medical purposes.

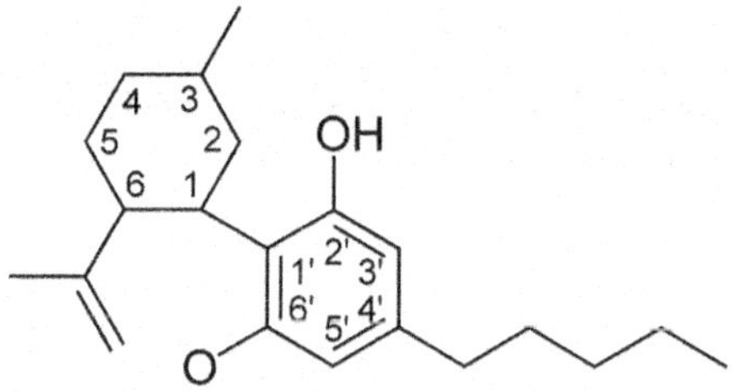

Chemical structure of a CBD-type cannabinoid

Some consumers of cannabis have the privilege of knowing which strains they are using. They understand that a certain strain may be better to relieve muscle pain after sporting activities, whereas another strain might have less of an effect on the body (often referred to as a body-stone) but give a more uplifting cerebral high. Some users may even know which strain they prefer for which specific activity or mood. More advanced aficionados with a connection to growers or an interest in cannabis biology will also know that a high coming from a plant that has been harvested too late (when most of the pistils of the plant have turned amber) will probably make them more sleepy and give a more disorientating, disruptive effect than a high coming from a plant of the same strain that has been harvested earlier, at the optimum point of the cycle.

A New Experiential Database for Scientific Research

Many marijuana users become aficionados because they are beginning to understand that marijuana holds great potential for them, a potential that can best be fulfilled if they know which strain to choose under which conditions. They have a vested interest in learning how different marijuana strains and variations in growing conditions bring about the various enhancements they enjoy so much. In a legal market, users would be better informed about which type of marijuana they are consuming. Standard product controls would ensure they could rely on the consistent quality and purity of the strain varieties they buy. I'm sure that thousands of those who value marijuana, including medical marijuana users, would be happy to contribute reports of their experiences to a scientifically structured database, possibly in the form of a website, requesting specific information about the effects of different strains on users.

We know that the end of marijuana prohibition would both save and generate huge amounts of money for governments. A scientific project asking for feedback on experiences would be comparatively cheap to finance. Such a website would be extremely effective as a resource for the study of the mind-altering properties of cannabis as well as its applications as a medicine for all kinds of purposes. Thousands of detailed anecdotal reports from knowledgeable users would help to inform scientific research tremendously, leading the way to a better understanding of which combinations of cannabinoids in marijuana are actually causing which effects on mind and body. The legalization of marijuana would make it possible for us to draw upon far more, and far better informed, anecdotal evidence. We could also ask contributors for more personal information, helping to better categorize their reports based on their personal histories and experience. This project would boost our understanding and research of cannabinoids, their medical applications and their potential for all kinds of enhancements, as well as their risks. Given the many medicinal and other uses of marijuana we already know, this WIKI marijuana research project could help millions, if not hundreds of millions of people around the world.

Carl Sagan, Cannabis, and the Right Brain Hemisphere

"What it comes down to is that modern society discriminates against the right hemisphere."

Roger Wolcott Sperry, 1913-1994, neuropsychologist and neurobiologist

Carl Sagan with the Viking lander that would land on Mars

In 1971, Harvard Prof. Lester Grinspoon published his milestone book "Marijuana Reconsidered", in which he featured an essay "Mr. X" by his best friend, the famous astronomer and science popularizer Carl Sagan. [151] In his essay, Sagan reports that for him, the marijuana high led to various cognitive enhancements, including the enhancement of cognitive abilities such as enhanced episodic memory, pattern recognition, creativity, and the ability to produce insights. More than 40 years later, Sagan's essay "Mr. X"

[151] Carl Sagan, „Mr. X", in: Lester Grinspoon (1971), *Marijuana Reconsidered*, Harvard University Press, Cambridge, Massachusetts, p.123 - 130. Sagan's essay was published anonymously, Grinspoon revealed the identity of his author only years after Sagan's untimely death.

is still one of the most illuminating accounts on the positive mind-altering potential of the marijuana high. Many other cannabis users before and after Sagan have described similar mind enhancements during a high. My own experiences with the marijuana high have been very similar, and my research aims to explain how good quality marijuana can induce insights and many other cognitive enhancements, when used by skilled users.[152]

Sagan's Hypothesis

Sagan was so excited about the cognitive enhancements he experienced with marijuana that he used it on a regular basis to come to insights for his work. His biographer Keay Davidson wrote that when Lester Grinspoon received unusual high-quality marijuana from an admirer by mail, he shared the joints with his friend Sagan and his wife Ann Druyan. *"Afterward, Sagan said, 'Lester, I know you have only got one left, but could I have it? I've got serious work to do tomorrow and I could really use it.'"*[153]

Sagan was famous – or, for many of his academic colleagues, infamous – for making brilliant but daring speculations – not only in his field of astronomy, but also in other scientific fields. As his biographer Davidson remarks, Sagan for instance wrongly predicted the existence of complex organic molecules on the moon; but, Davidson adds, *"(a)s it turned out later, complex organic molecules pervade much of the outer solar system and beyond."*[154]

In his Pulitzer Prize-winning book "*The Dragons of Eden – Speculations on the Evolution of Human Intelligence*"(1977), Sagan summarized what cognitive and evolutionary scientists had to say about the evolution of the human mind and presented many of his own brilliant speculative ideas on the nature and evolution of human intelligence. He also introduced a speculative thesis about the effects of marijuana on the human brain. Sagan

---

[152] See Sebastián Marincolo (2010), *High. Insights on Marijuana*, Dog Ear Publishing Indianapolis, Indiana.

[153] Keay Davidson, (1999), *Carl Sagan. A Life.*, John Wiley & Sons, Inc., New York, p.214.

[154] *Ibid.*, p.213.

started from an analogy made by psychologist Robert Ornstein:

"He {Ornstein} suggests that our awareness of right hemisphere function is a little like our ability to see stars in the daytime. The sun is so bright that the stars are invisible, despite the fact that they are just as present as they are in the daytime as at night. When the sun sets, we are able to perceive the stars. In the same way, the brilliance of our most recent evolutionary accretion, the verbal abilities of the left hemisphere, obscures our awareness of the functions of the intuitive right hemisphere, which in our ancestor must have been the principal means of perceiving the world."[155]

In a now famous footnote to this paragraph Sagan formulates his hypothesis on how a marijuana high could affect thinking:

"Marijuana is often described as improving our appreciation of and abilities in music, dance, art, pattern and sign recognition and our sensitivity to nonverbal communication. To the best of my knowledge, it is never reported as improving our ability to read and comprehend Ludwig Wittgenstein or Immanuel Kant; to calculate the stresses of bridges; or to compute Laplace transformation. (...) I wonder if, rather than enhancing anything, the cannabinols (the active ingredient in marijuana) simply suppress the left hemisphere and permit the stars to come out. This may also be the objective of the meditative states of many Oriental regions."[156]

Enhancements and Suppression

Let me distinguish two aspects of Sagan's hypothesis: First, there is his observation that a marijuana high leads to a style of thinking which cognitive scientists consider to be predominantly based in the right hemisphere. Second, Sagan speculates that a marijuana high might suppress left hemisphere functions and, therefore, leads to what we could call "right hemisphere thinking". Sagan's hypothesis was based on his knowledge of what current science had to say about the human brain and the different

---

[155] Carl Sagan (1977), *The Dragons of Eden. Speculations on the Evolution of Human Intelligence*, Random House Publishing Group, New York, p.177

[156] *Ibid.*

style of cognition in the left and right brain hemispheres, as well as on his own experiences with marijuana.

Sagan's Claim Reconsidered

During his lifetime, Sagan did not reveal to the public that he was a user of marijuana. In isolation, Sagan's footnote about marijuana and its possible effect on the brain hemispheres may seem to be a spontaneous speculation out of nowhere. Yet, when reading Sagan's essay "Mr. X", it becomes clear that Sagan's experiential basis for such a claim was actually quite broad. In his essay, he describes in some detail not only the enhancements outlined in his footnote, but he also mentions other cognitive enhancements, such an enhanced ability to remember past events and to obtain deep insights. Also, Sagan mentions that during a high, he also experienced enhanced…

"(…) perceptions of real people, a vastly enhanced sensitivity to facial expressions, intonations, and a choice of words which sometimes yields a rapport so close it's as if two people are reading each other's minds."[157]

Sagan already knew that most of the cognitive enhancements he had experienced during a high concern functions which cognitive science at the time found to be predominantly located in the right brain hemisphere; he had described the research concerning split brain patients and more on the pages preceding his hypothesis in "The Dragons of Eden".

So, Sagan knew this research and had made his own experiences with cognitive enhancements during a marijuana high. Furthermore, he was also well acquainted with his friend Lester Grinspoon's research about marijuana as a medicine and about the mind-enhancing potential of marijuana. As Lester Grinspoon told me in private conversation, Carl Sagan had carefully read and commented upon his manuscript of *Marijuana Reconsidered*, in which Grinspoon extensively featured reports of other marijuana or hashish users who had described similar mind enhancements.

---

[157] Carl Sagan, „Mr. X", in: Lester Grinspoon (1971), *Marijuana Reconsidered*, Harvard University Press, Cambridge, Massachusetts, p.129.

To summarize, then, Sagan's hypothesis certainly didn't come out of nowhere. On the contrary, his knowledge about the differences in right and left hemisphere cognition style was quite impressive, and he knew a lot about the altered style of cognition during a high from his own experiences and from countless user's reports he knew from Grinspoon's work.

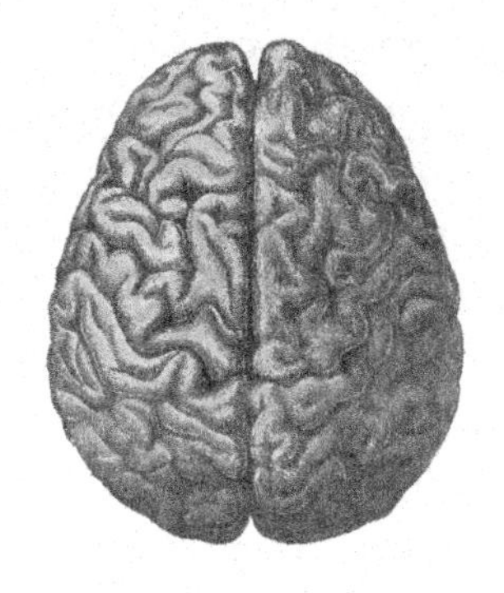

The left and the right brain hemisphere.

Now, what can we say about Sagan's hypothesis almost 40 years later, with all the progress in the mind sciences on the nature of cognition in the respective brain hemispheres and with our knowledge of the endocannabinoid system and its many functions in cognition? Was Sagan on the right track?

Tragically, Sagan died in 1996, too early to witness the revolutionary discovery of the endocannabinoid system which first came to light around that time. Since then, we have learned about the amazing range of its physiological and cognitive functions. So, can we actually find evidence for or against Sagan's hypothesis for instance looking at the distribution of endocannabinoid receptors (especially the CB-1 receptor, on which the exogenous THC acts)?

Sagan's Thesis and the Endocannabinoid System

As far as I can see, there seems to be no extensive research concerning the subject of lateralization and the role of endocannabinoid signaling for higher cognitive functions. One often cited study found an increased blood flow in parts of the right hemisphere during a high[158], but another study states that *"relatively high concentrations of cannabinoid receptors were consistently seen in cortical regions of the left (dominant) hemisphere, known to be associated with verbal language functions.*"[159]

[158] Roy Mathew *et al.* (1997) , „Marijuana intoxication and brain activation in marijuana smokers", *Life Sci.* 1997;60(23): 2075-89.

[159] Glass, M., Dragunov, M., Faull, RL (1997) „Cannabinoid receptors in the human

It seems much too early to draw conclusions from brain imaging studies to evaluate Sagan's hypothesis. Even better data on the concentrations of endocannabinoid receptors in the respective brain hemispheres will not directly lead to resolving the issue; Marsicano and Kuhner remind us that *"sometimes the endocannabinoid system appears to be functionally important in regions or cell types where the density of CB1 receptor is relatively low {e.g. control of pain perceptions in the brainstem}."*[160] As far as I can see, we haven't yet begun to understand how the endocannbinoids are involved in affecting different higher cognitive functions based in the left or right brain hemisphere.

Recent Neuroscience and Reports about the Marijuana High

Some more support for Sagan's thesis seems to come from two other sources. First, Sagan's description of the cognitive enhancements during a marijuana high have been described in detail by many more cannabis users.[161] The most impressive collection of recent anecdotal reports and essays about those enhancements comes from Sagan's best friend Lester Grinspoon and can be found on his website *marijuana-uses.com*. Many of the enhancements reported by marijuana users seem to be based on a shift towards a more right-hemisphere cognition style during a high.

Second, if we look at what recent neuroscience has to say about the differences in the cognitive functions and processing styles in the two brain hemispheres, Sagan still seems to have a point. From what we know now, the right hemisphere plays an important role for the cognitive processes named by Carl Sagan as enhanced during a high, as well as for many other enhancements described by other marijuana users. In his book *The Master*

---

brain: a detailed anatomical and quantitative autoradiographic study in the fetal, neonatal and adult human brain." *Neuroscience.* 1997 Mar; 77(2): p.299-318.

[160] Marsicano, G., and Kuner, R. (2008), „Distribution of CB1 Cannabinoid Receptors in the Nervous System", in: Attila Köfalfi (ed.) (2008), *Cannabinoids and the Brain*, Springer Science and Business Media, New York, p.164.

[161] Compare Sebastián Marincolo (2010), *High. Insights on Marijuana*, Dog Ear Publishing Indianapolis, Indiana.

and his Emissary. The Divided Brain and the Making of the Modern World, (2009), psychiatrist Iain McGilchrist gives a survey of the current scientific knowledge about the different styles of cognition in the left and the right brain hemisphere emerging from stroke cases in one hemisphere, split brain patient cases, and newer brain imaging studies. According to McGilchrist, the right hemisphere is predominately involved in our ability to remember personal events (episodic memory), is responsible for associations between widely different concepts and ideas, complex pattern recognition, creative problem solving, insights, the appreciation of humor, the understanding of metaphors, self-awareness, the empathic understanding of others, the processing of words describing the mind, and the interpretation of emotional expression in faces, in intonation and verbal implications. Also, the right hemisphere seems to be crucially involved in the interpretation of non-verbal communication and the perception of music.

For the sake of brevity, I will leave it at this incomplete list, but actually, McGilchrist describes more cognitive functions as right hemisphere-based, functions which all have been consistently reported by countless users to be enhanced during a high.

The Effect of Marijuana on Attention

In general, then, it seems that research of the last 40 years in the neurosciences lend some more support Sagan's thesis that marijuana leads to an enhancement of right hemisphere-based cognitive abilities. One interesting puzzle, though, is one of the fundamental effects of marijuana on attention. When high, users often get completely absorbed by the taste of ice cream, by an intense stream of memories or ideas, or by the sensation of a kiss.

In other words, a high seems to cause a strong selective focus on whatever we attend to. According to McGilchrist, however, focused attention is not a cognitive function performed primarily in the right hemisphere.

On the contrary, he summarizes the current research as saying that *"(…) the right hemisphere is responsible for every type of attention except focused attention."*[162]

Evaluating Sagan's Hypothesis

So, when it comes to attention, at least one of the typical cognitive changes during a high does not seem to come from an enhancement of processes in the right hemisphere. Generally, then, while Sagan seems to have been basically on the right track, not all of the cognitive enhanced functions during a high seem to be predominately based in right hemisphere activities. We will have to wait for more research in this field to see exactly how cannabis affects cognitive activities in the left and right brain hemisphere.

What about Sagan's hypothesis that a high could suppress left hemisphere function and, thus, 'bring out the stars' and allow for an enhanced right hemisphere activity? As far as I can see, we are far from being able to tell whether the mind enhancements during a high observed by so many users come from a direct enhancement of certain cognitive functions, or whether they arise from a suppression of some left hemisphere activities.

Laggar falcon in flight

According to McGilchrist, the left and the right hemisphere are in a constant battle for control. In order to help us to survive, we and other animals need two conflicting systems of attention. He explains this point with the example of birds: to pick up seeds for food, a bird needs to

[162] McGilchrist, Iain (2009), *The Master and His Emissary, The Divided Brain and the Making of the Modern World*, Yale University Press, New Haven and London, p.39.

narrowly focus on the seeds on the ground to motor control and coordinate food in-take (left hemisphere function); but in order to survive, the bird has to get distracted by a predator like a falcon in the fringe of its perception. So there must be another type of attention drawing it to unusual new sensations (right hemisphere). Only the interplay of these competing attention systems located in the two brain hemispheres allows birds and other animals to survive. Clearly, then, an enhancement of cognitive processes in one hemisphere could come from the suppression or weakening of cognitive functions in the other hemisphere.

We still have to await further research to come to a better understanding on how consumed cannabinoids affect the endocannabinoid system and, generally, its role in higher cognition. Yet, almost 40 years after Sagan's hypothesis, I think we can still say that he was on an interesting track. As far as I can see, we will only come to a better understanding of the nature of the high and its effects on attention, memory, pattern recognition, creativity, empathy, or insights, once we begin to research how endocannabinoids and consumed marijuana affects various cognitive processes in the left and the right brain hemisphere.

The Historical-Sociological Perspective

Sagan's hypothesis does not only have interesting implications for individual marijuana users. In his book *The Master and his Emissary. The Divided Brain and the Making of the Modern World*, Iain McGilchrist argues in detail that in various periods of history there were general shifts in our (Western) society toward more left or right hemisphere cognition styles which can be detected in art, science, philosophy, literature and politics:

"My thesis is that for us as human beings there are two fundamentally opposed realities, two different modes of experience; that each is of ultimate importance in bringing about the recognizably human world; and that their difference is rooted in the bihemispheric structure of the brain. It follows that the hemispheres need to co-operate, but I believe that they are in fact involved in a sort of power struggle, and that this explains many aspects of contemporary

Western culture.[163] *(...) I believe that over time (...) the balance of power has shifted where it cannot afford to go – further and further towards the part-world created by the left hemisphere."*[164]

For McGilchrist, this means that in our society we now predominantly rely on left hemispheric thinking; which leans towards abstraction, is more rational, mechanistic, less empathic, less open to the richness of holistic experience, and "yields clarity and power to manipulate things that are known, fixed, static, isolated, decontextualized, explicit, disembodied, general in nature, but ultimately lifeless."[165]

The right hemisphere, by contrast, represents the world to us as *"individual, changing, evolving, interconnected, implicit, incarnate, living beings within the context of the lived world, but in the nature of things never fully graspable, always imperfectly known – and to this world it exists in a relationship of care."*[166]

Just like Sagan, McGilchrist does not simply speak in favor of right hemisphere thinking, but reminds us that sane human cognition comes out of a balanced interplay between the two hemispheres. If McGilchrist is right, then what we dramatically need in our society is to come back from our imbalanced left-hemispheric thinking by using mindfulness or techniques like meditation.[167] If Sagan is right with his general hypothesis that a marijuana high actually does cause a more right-hemispheric style of thinking, then it is easy to see how skilled and knowledgeable marijuana use could lead to positive changes in the mindset and thinking of our society as a whole. If we look back in recent history, we have already seen these changes happening and leading to vastly important positive cultural changes, from the early evolution of jazz to the peace movement of the Beat Generation.

---

[163] *Ibid.*, p.3.
[164] *Ibid.*, p.6.
[165] *Ibid.*, p.174.
[166] *Ibid.*, p.174.
[167] Margaret Emory (2012) „Dr. Iain McGilchrist on the Divided Brain, Q&A with Iain McGilchrist", *BrainWorld*, May, 8 2012, *http://brainworldmagazine.com/what-at-any-one-moment-is-governing-our-actions/*

From this perspective, more widespread skilled and respectful marijuana use might well play a defining role in elementary changes in our society yet to come; changes, which could possibly help to save the world we live in.

Appendix Off to New Shores

Vaporizer Highs

What characterizes a vaporizer high? How much of a difference is there between a vaporizer high and the high from a joint or a bong? How different are various vaporizers on the market as to the high they produce? How much does the temperature setting on a vaporizer matter to the quality and character of a cannabis high?

A Clearer High

There are many discussions about these and other questions concerning the vaporizer high. After reading various online threads from a number of cannabis websites I would conclude that there is at least one thing almost everybody vaporizer user agrees on: the vaporizer high is much 'clearer' than the high coming from burnt cannabis smoked either from a bong, a pipe, or from a joint. Going from my own experiences and from reports of many other users, a 'clear' high is a high that leaves you cognitively more functional; there are less short-term memory disruptions; you are less likely to lose the thread when talking about a certain subject; you feel less tired, less disoriented and confused. One reason for this is probably due to the absence of the various toxins created by burning cannabis at high temperatures – in a burning joint, temperatures are around 700-1100 °F. When cannabis is heated beyond 392 °F, some unwanted substances are produced:

"(...) traceable amounts of benzene are found in the vapor mist. Benzene contributes to couch lock {a sedative effect that makes you want to stay on your couch}(...) [168]

But the mental clarity of the vaporizer high compared to highs from burning cannabis cannot only be explained by the absence of some toxins. A cannabis plant contains more than 80 cannabinoids. It also contains more

---

[168] Martin, Alexander (2012), "Tailoring Your High: Compounds in Cannabis, Properties and Boiling Points", http://www.weedist.com/2012/07/tailoring-high-compounds-in-cannabis-properties-boiling-points/

than 200 terpenes and dozens of flavonoids, which are responsible not only for the distinct aroma of a strain, but are now also known to have an important influence on the high.

According to microbiologist and biochemist, Kathleen O'Dea,

> *"(c)annabis produces over 200 different terpenes in the plant's resin glands and each strain has a unique terpene "signature". Nearly 20 percent of the total oil produced by the plant's resin glands are terpenes and up to 30 percent of marijuana smoke is made of terpenes. In fact, it is the terpenes that are "sniffed out" by trained drug dogs! No other plant on the planet has the potential to create the range of aromas and flavors known to Cannabis. The most important aspect of terpene production is the fact that terpenes can interact synergistically with other compounds in the plant resulting in a kaleidoscope of healing effects."*[169]

Female cannabis plant in the flowering stage. Photo © Sebastián Marincolo 2012

If we want to better understand the character of vaporizer highs, we first need to take a look at the different boiling points of the relevant cannabinoids, terpenes and flavonoids. Let us look at the cannabinoids first. THC (delta-9-tetrahydro-cannabinol), which is said to give you a more heady, cerebral, energetic high, boils already at 314.6 °F. CBN (cannabinol, an oxidation breakdown product) which is known to be a sedative and to generate a more confusing, disorienting high, boils at 365 °F. CBD (cannabidiol), now becoming

[169] Rutherford, Susan "The Science of Cannabis– Kathleen O'Dea", January 7, blog post http://www.naturesalternativepdx.com/science-cannabis-kathleen-odea/

more renowned for its medicinal value for various purposes (anxiolytic, analgesic, antipsychotic, antispasmodic, etc.) boils between 320-356 °F. The cannabinoid THCV (tetrahydrocannabivarin), which is known to act as a euphoriant as well as being analgesic, boils at 428 °F. For the sake of brevity I'll leave it at this short list; but there are other cannabinoids that we know have an influence on the high.

Now, given that vaporizers can be set to different temperatures ranging from as low as 266 °F to 392 °F and more, it should be clear that they can produce a range of markedly different highs at various different temperatures. Many experienced vaporizers users agree with the following statement that lower vaporizer temperatures leave you with a more cerebral, uplifting, euphoriant high:

"(...) when you have a precision vaporizer, you can set the temperature so it only produces THC. The magic is, THC vaporizes after you hit about 305 degrees Fahrenheit but other cannabinoids don't vaporize (and won't be delivered into your lungs) unless you set the vaporizer to approximately 380-415 degrees Fahrenheit. So no matter what marijuana strain you have, if you only want to experience THC effects, you dial 305-320° F. You only get vaporized THC, and you avoid CBD and CBN completely."[170]

Terpenes and Flavonoids

Many plants and some insects produce terpenoids for a whole variety of reasons. When ingested by animals, terpenes can be anti-oxidant, analgesic, anti-microbial, anti-carcinogen, muscle relaxant, anti-depressant, anti anxiety, sedative, and psychoactive in various ways, and have various other effects. Plants need to defend themselves against pests and herbivores but some also have to rely on animals for reproductive purposes:

"Many plants that rely on animals to disperse their seeds must limit feeding rather than kill the animal directly because it might lose a valuable seed carrier.

---

[170] Davis, Steve (2014) „Marihuana Vaporizer: Your Way to Incredible New Marihuana Highs“, http://bigbudsmag.com/grow/article/Marihuana-vaporizers-thc-cbd-cbn-vaping

Lipids (...) perform this function. The most interesting lipids are terpenes, which have a strong aromatic flavor. For example, myristicin, found in nutmeg and many other spices, prevents animals from seasoning their diet too heavily with those plants. If taken in sufficient quantities, myristicin causes dizziness and loss of motor coordination."[171]

Macro photography of a cannabis leaf with the mushroom shaped trichomes, where the plant produces its cannabinoids, terpenes, and flavonoids. Photo © Sebastián Marincolo 2013

In much the same way as the various cannabinoids, terpenes and flavonoids that have been developed by plants in order to improve their interaction with animals. These chemical compounds are geared to interact with our biochemistry and to alter biochemical processes in many ways. Therefore, they can be used for a whole range of medical purposes and have many diverse influences on mind and cognition. The distinctive terpene profile of a certain strain contributes immensely to the high coming from that strain:

---

[171] Siegel, Ronald K. (1989, 2005), *Intoxication, The Universal Drive For Mind Altering Substances,* Park Street Press, Rochester, Vermont, p.29

"Terpenes also bind to these {endocannbinoid} receptor sites and affect their chemical output. They can also modify how much THC passes through the blood-brain barrier. Their hand of influence even reaches to neurotransmitters like dopamine and serotonin by altering their rate of production and destruction, their movement, and availability of receptors.

The effects these mechanisms produce vary from terpene to terpene; some are especially successful in relieving stress, while others promote focus and acuity. Myrcene, for example, induces sleep whereas limonene elevates mood. There are also effects that are imperceptible, like the gastroprotective properties of Caryophyllene."172

The boiling points of terpenoids (when dried and cured, terpenes turn into terpenoids) are also distributed over a whole range of temperatures. Myrcene, the most prevalent terpene found in today's marijuana has clove-like, earthy, citrus, mango and minty nuances [173], boils between 331-334 ºF and is said to be sedative, hypnotic, analgesic as well as contain anti-inflammatory and muscle relaxant properties. According to marijuana expert Ethan Russo, myrcene contributes heavily to the couch-lock "stoned" effect.[174] D-limonene boils at 351 ºF and has hints of citrus fruits, rosemary, juniper and peppermint, is anti-bacterial, repulsive to predators and is found in many species of fruits and flowers. It can also be anti-carcinogenic and an anti-depressant and has been used to dissolve gallstones. The sedative 945-terpineol boils at only 423-424 ºF and has hints of lilac, citrus, apple blossoms and lime. The flavonoid apigenin boils at 352 ºF and is believed to have anxiolytic and anti-inflammatory properties.

This short list of terpenes and their characteristics already illustrates that we should not look just at dozens of cannabinoids, but also at a whole variety of terpenes as well as flavonoids and their distinctive boiling points if we

[172] Rahn, Bailey (2014) "Terpenes: the Flavors of Cannabis Aromatherapy", http://www.leafly.com/news/cannabis-101/terpenes-the-flavors-of-cannabis-aromatherapy, 2/12/2014.

[173] Vogeler, Josh (2014), "Terpenes and Terpenoids in Cannabis", http://terpenes.weebly.com.

[174] Lee, Martin A. (2013), "Talking Terpenes," *High Times*, 4/8 2013, http://www.high-times.com/read/talking-terpenes.

want to understand the distinctive highs coming from various strains vaporized at a certain temperature.

Let me add that for aficionados, there is another good reason to prefer a vaporizer to burning cannabis in a joint or in a device like a bong. Vaporizers do not burn terpenes and can therefore bring out the full aromatic bouquet of a strain. You will not only get a high different from that of burned cannabis because you get a more favorable cannabinoid/terpene/flavonoid profile without the toxins. Your high will also be positively affected by a distinctive sensual aromatic experience. When you drink a 30-year-old Port Ellen whiskey, the alcohol does not only affect your consciousness as by a psychoactive substance. The complex aromatic experience opens up your mind, makes you more sensitive and allows you to go on a sensuous trip. Vaporizer users can bring more of this aficionado component to their consumption, especially when vaporizing marijuana at lower temperatures. Also, as we have seen, there is a connection between the smell that comes from the terpenes (cannabinoids are odorless) and some medical and psychoactive characteristics of the high; for some patients and others, the odor and taste of cannabis vapor coming from a vaporizer will help to identify their preferred strains and to check if the plant material they obtain has been produced, stored and cured correctly.

Vaporizer technology and user experiences

If we want to evaluate experiential reports about vaporizers we have to look very closely at what kind of vaporizers have been used to generate a high. As Markus Storz, the inventor of the German Volcano vaporizer, probably the best known precision vaporizer in the world, explained to me in a personal interview:

"Vaporizers which heat up only the chamber, but do not heat up the incoming air to vaporizing temperature cannot consistently heat up the plant material."

Many experimental reports from users of vaporizers with an inferior heating apparatus have to be considered with care. While most vaporizers may still deliver a much cleaner high than one from a joint or bong, users of many inferior vaporizers do not get the precision temperature control to really let them join in a serious discussion about the differences of a high coming from a certain strain produced with a vaporizer at 320 °F or 356 °F.

And there are other factors we have to bear in mind when evaluating personal reports about the vaporizer high from users. Many users have smoked joints for a long time before using vaporizers, usually made of low-quality black market cannabis mixed with inferior tobacco. They tend to equate the 'real' high with the resulting effects of burned, inferior cannabis and bad tobacco – which are usually more disruptive, disorienting, sedating and often messing with their short-term memory. Some users are initially disappointed by the vaporizer high from pure cannabis because they miss a certain nicotine kick from the tobacco, or because they have actually been seeking out a 'mind-crippling' effect which helps with sedation and forgetting the trials and tribulations of their day. I have talked to many users who told me they seek a state of mind in which they simply get a body stone relaxation coupled with a high that heavily interferes with their short-term memory, so that they can get relief from their daily stresses. A clearer high coming from a precision vaporizer set to a lower temperature does not give them what they want. Many of these users do not even recognize the altered state of mind coming from a vaporizer as a real high, because they are not used to being so cognitively functional.

When we look at discussions about the vaporizer high in various internet forums we also need to keep in mind that many users receive their cannabis from the black market and, therefore, often do not know exactly what strain they are buying, under which conditions it has been produced and stored and what the cannabinoid profile of their cannabis really is. Briefly then, many of the anecdotal reports about vaporizer highs and the generalizations about how vaporizers affect the high have to be considered with care. Many opinions are based on consumers using bad quality marijuana with vaporizers that do not allow for a precise temperature control, and many user reports are influenced by their bias from smoking marijuana laced with tobacco. This also explains the many contradictory reports from users.

Precision vaporizers and the future of cannabis research

When we look at the various cannabinoids and their boiling points, we can certainly make some rough predictions about the systematic influences of a vaporizer on the character of a high. For instance, temperatures higher than 365 °F will produce more CBN, which is known to produce a more sedative, confusing effect on consciousness. A vaporizer high will always strongly depend not just on a certain strain and its cannabinoid and terpene profile, but also on the exact temperature at which it is used. Only precision vaporizers like the Volcano in the hands of skilled users will help us to answer many questions and to come up with new questions about how a vaporizer can affect a high.

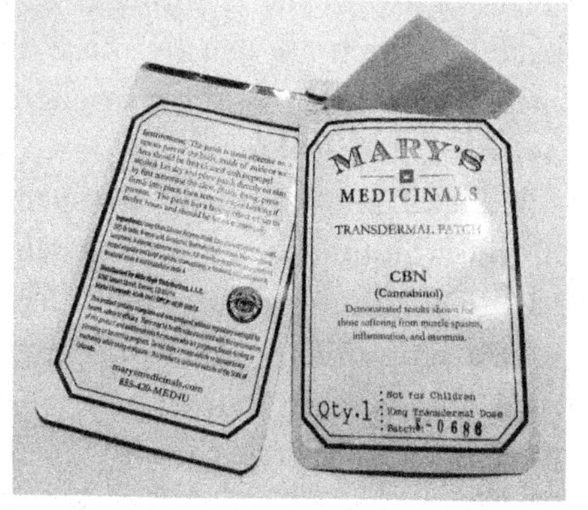

Cannabinol transdermal patches sold at dispensaries in Colorado, from Wikipedia commons

Using precision vaporizers to explore the differences and nuances in a marijuana high may sound to many like a somewhat futile exercise to answer nerdy questions of a small group of geeky aficionados. But make no mistake: answering these questions will help millions of people to make better use of the medical potential of marijuana; and millions will find out how to use marijuana in a more meaningful and inspiring way. Countless people around the world already use the marijuana high to better remember long gone events, to work creatively, or to find new patterns in music or art. They use it to better appreciate nature, to get in touch with their feelings and have a better introspective access to themselves, to enhance their empathic understanding of others, to make love, to generate great and live-changing insights, and to develop personally growth.[175] Many of those users are on a voyage, exploring a new world and the use of marijuana can be crucially important for their lives – and the lives of everybody around them. They and their societies will profit greatly from the advance in our understanding as to how the various cannabinoids, terpenes and flavonoids affect a high.

---

[175] Compare Grinspoon, Lester (ed.) (2014), marijuana-uses.com.

Precision vaporizers are magnificent tools for researching the psychoactive and bodily properties of the various cannabinoids, terpenes and flavonoids, and we are only at the very beginning when it comes to understanding these substances and their full inspirational and medical potential. Just as importantly, the effects of cannabinoids, terpenes and flavonoids are already known to be synergistic, which means that we cannot simply study the effects of them individually as isolated chemical compounds. Vaporizers with their temperature control can help us to study the compounds in various natural strains of marijuana as they work synergistically on our mind and body.

Exploring New Dimensions

To sum up, precision vaporizers offer new dimensions in exploring the mind-altering potential of cannabis. A marijuana high is the result of the effect of dozens of cannabinoids, terpenes and flavonoids, which all have different boiling points and therefore vaporize at different temperatures. Many of the generalizations about vaporizer highs that can be found in forums are wrong and often contradictory; others, such as the claim that the vaporizer high offers a much "clearer", less confusing and less sedated experience make sense not only regarding the many users who endorse this view, but also on the basis of what we know so far about the psychoactive properties of some cannabinoids, terpenes and their various boiling points.

More and more users are picking up the habit of using precision vaporizers and so, hopefully, our knowledge about the psychoactive and medical role of the various cannabinoids, terpenes and flavonoids will expand. There is already much knowledge concerning these questions coming from the medical dispensaries in the U.S. and their patients. I hope that many inspirational users will soon be in a better position to access a much higher quality of marijuana and to enjoy its effects in precision vaporizers. Their feedback and that of medical marijuana users to medical and other professionals will be crucial for the future research of the potential of marijuana. Vaporizers will play a fundamental role in this journey, a journey that will essentially help millions of marijuana users in the future.

Many animals have a very intelligent use of the plants around them. The Australian Koala bear, for instance, eats only select types of eucalyptus. Some aromatic oils of the eucalyptus digest from their stomachs through their skins to protect them against ectoparasites, while other oils "*decrease blood pressure, lower body temperatures, and relax their muscles.*"[176] Usually, Koalas eat the mature leaves for these purposes in hot climates, but in cold climates they eat younger leaves containing phellandrene, which increases body temperature.

Koalas, then, instinctively know their eucalyptus well. It is time for us cannabis users to use our much larger brains to behave at least as smart as they do.

---

[176] Siegel, Ronald K. (1989/2005), *The Universal Drive for Mind Altering Substances*, Park Street Press, Rochester, Vermont, p.43.

Thanks

I would like to thank Jeroen Roeleveld for his trust and for giving me the opportunity to write an expert blog for the cannabis seed bank Sensi Seeds. The essays in this book are 'enhanced' versions of the original essays published for this blog from 2012 to 2015. Without this support the necessary research for this book would not have been possible.

Very generous support came from Carl Doherty, who helped me extensively with editing and proofreading for no costs, even though I offered. Eternal thanks, Carl. Excellent job! The remaining mistakes are definitely on me.

This book was published with the help of an Indiegogo crowdfunding campaign. I thank my anonymous supporters as well as some who chose to openly support me: thanks Lieven D'hont, Paul Aguilar, and Shari L. Mathieu.

Thanks, Joe Dolce, for contributing a foreword and for sharing so much of your knowledge about the cannabis world with me in our conversations.

Last but not least, thanks again to Andy Smith for the magnificent cover design.

Dr. phil. Sebastián Marincolo

Photo © Tom Lichtenbergh 2014

is a former student of the philosophers Manfred Frank, Gianfranco Soldati, William G. Lycan, Simon Blackburn, and Dorit Bar-On. His research focuses on the philosophy of mind, neurocognition, and on altered states of mind. He has received several academic grants and fellowships, including a Fulbright grant and a fellowship from the German Academic Exchange Service (DAAD).

Marincolo has published various articles on the marijuana high, co-edited *bewusstseinserweiterungen ("mind expansions")*, an issue of the German internetzine *parapluie.de*, and published two books on the marijuana high: the study *High. Insights on Marijuana* (Dogear Publishing, Indiana, USA, 2010), and an essay collection, *High. Das positive Potential von Marijuana* (in German, Klett Cotta/Tropen, Stuttgart, Germany, 2013), which contains his macro art series "The Art of Cannabis". The new essays in the present collection first appeared – most of them in shorter versions – in five languages on his expert blog for the renowned Dutch cannabis seed bank *Sensi Seeds,* the official provider of medical marijuana in the Netherlands, between 2013-15. Marincolo also worked with marijuana expert Harvard Associate Prof. Emeritus Lester Grinspoon on a book project. For more than five years he was a creative director and consultant for one of the biggest foundations in Germany and has more than 25 years of experience as a freelance photographer specializing on documentary, art, travel, and macro-photography. His art photography from New York City, Rio de Janeiro, Bali and other places has been shown in various exhibitions and art galleries in Germany and the U.S. He is currently living in Stuttgart, Germany, and works as a freelance writer, creative consultant and photographer.

Personal home page and blog:
www.sebastianmarincolo.de
www.marijuana-insights.com